ENVIRONMENT, GROWTH AND DEVELOPMENT

Is sustainable development the answer to environmental decline and development failure? In 1987 the Brundtland Commission concluded that sustainable development would integrate environmental concerns into mainstream policies, shifting the focus from weak and peripheral environmental management to the socio-economic policy sources of environmental impacts. The 1992 Earth Summit confirmed the need to integrate environmental protection into the development process. It also requested that all countries establish programmes of integrated environmental and economic accounting for use in sustainable development planning and policies. An increasing number of developing and developed countries are now implementing 'green accounting' to formulate national policies and strategies of sustainable economic growth and development.

Environment, Growth and Development offers a unique analysis of sustainable economic growth and development based on operational variables derived from the new systems of 'green accounting'. A complete revision and expansion of *Environment and Development*, this book offers a new focus on macroeconomic aspects, comparing the 'goods' of economic production and consumption with the 'bads' of losses of natural resources and environmental quality. Beyond economics, ways of evaluating social, cultural, aesthetic or ethical issues are also proposed.

Focusing on operational, quantifiable concepts and methods, the book systematically links the different policies, strategies and programmes of growth and development to advance an integrative policy framework for sustainable development at local, national and international levels in both developing and industrialized countries.

Peter Bartelmus is the Officer-in-Charge of the Environment and Energy Statistics Branch of the United Nations Statistical Division, New York.

ENVIRONMENT, GROWTH AND DEVELOPMENT

The concepts and strategies of sustainability

Peter Bartelmus

London and New York

First published 1994
by Routledge
11 New Fetter Lane, London EC4P 4EE

Simultaneously published in the USA and Canada
by Routledge
29 West 35th Street, New York, NY 10001

Typeset in Garamond by J&L Composition Ltd, Filey, North Yorkshire
Printed and bound in Great Britain by T.J. Press (Padstow) Ltd, Padstow, Cornwall

British Library Cataloguing in Publication Data
A catalogue record for this book is available from the British Library

Library of Congress Cataloging in Publication Data
Bartelmus, Peter.
Environment, growth and development: the concepts and strategies
of sustainability/Peter Bartelmus.
p.   cm.
Includes bibliographical references and index.
1. Economic development—Environmental aspects.   2. Sustainable
development.   3. Environmental policy—Economic aspects.   I. Title.
HD75.6.B373   1994
333.7—dc20   93–34234

ISBN 0–415–10669–9 (hbk)
ISBN 0–415–08484–9 (pbk)

To Liza

CONTENTS

List of tables, figures and text boxes   ix
Foreword   xii
Preface   xiv
Acknowledgements   xvii
Abbreviations   xviii

1 WHAT ON EARTH IS WRONG?   1
  1.1 Development and environment: from global discussion to
      global frustration   1
  1.2 Diagnosis: a planet at risk   8
  1.3 Evaluation: how bad is it?   26

2 ACCOUNTING FOR SUSTAINABLE DEVELOPMENT   31
  2.1 Environment and economic process   31
  2.2 A system for integrated environmental and economic
      accounting   34
  2.3 Pricing the priceless: the limits of monetary valuation   45
  2.4 Case-studies   53

3 ENVIRONMENT AND DEVELOPMENT – A MATTER
  OF SUSTAINABILITY?   60
  3.1 Sustainability criteria   60
  3.2 Definitions of sustainable growth and development   69
  3.3 Sustainability and optimality   75

4 PLANNING AND POLICIES I: SUSTAINABLE
  GROWTH AND STRUCTURAL CHANGE   79
  4.1 Policy framework   79
  4.2 Eco-variables in macroeconomics   85
  4.3 Cost internalization and structural change   98

5 PLANNING AND POLICIES II: SUSTAINABLE
  DEVELOPMENT   106
  5.1 Integrated planning and policies   106
  5.2 Grass-roots strategies: ecodevelopment   115

*5.3 Implementation: towards a national action plan for
sustainable development*   126

6 OUTLOOK: FROM NATIONAL TO GLOBAL
COMPACTS   140
6.1 Towards global sustainability   140
6.2 Earth Summit: an agenda for the twenty-first century   144
6.3 From national to global compacts   148

References   152
Index   161

TABLES, FIGURES AND TEXT BOXES

Tables

1.1 Country categories by level of growth and development   4
1.2 Environmental concerns of developing and industrialized
countries   13
1.3 Selected indicators of (non-)sustainability   14
2.1 Framework for integrated environmental and economic
accounting   37
2.2 Comparison of conventional and environmental accounting –
selected indicators: Mexico and PNG   56
2.3 Nature sector accounts for Chesapeaka – 1985   58
4.1 A framework for sustainable growth and development   82
4.2 Income and employment effects of environmental policy   95
4.3 Application of effluent charges in OECD member countries   104

Figures

1.1 The finite global ecosystem relative to the growing economic
subsystem   9
1.2 Projected man-made CO_2 emissions   18
1.3 Moderate, severe and extreme land degradation as a per cent
of vegetated land, 1945–90   20
1.4 Man-made emissions of common air pollutants   21
1.5 Marine fish catch as a per cent of estimated average sustainable
catch, 1987–9   22
2.1 Environmental–economic system   33
3.1 Sustainability and welfare generation   62
5.1 Man and ecosystem   117
5.2 Obergurgl simulation model of social, economic, demographic
and environmental interactions   123
6.1 The results of UNCED   146

Text boxes

1.1	Basic human objectives	2
1.2	International Development Strategies of the United Nations	3
1.3	New International Economic Order (NIEO)	7
1.4	Objectives of sustainable development	8
1.5	Facing up to change	11
1.6	Environmental issues	11
1.7	Poverty in developing countries	23
1.8	Balance sheet of human development	28
1.9	Economic–ecological interdependence	29
2.1	Classification of defensive expenditures	39
2.2	Natural assets, resources and resource accounting	41
2.3	Concepts of capital accumulation	43
2.4	Definitions of NDP and EDP	44
2.5	Valuation of natural resource depletion	47
2.6	Cost/benefit valuation of environmental impacts	49
2.7	Corporate environmental accounting	51
2.8	Net Economic Welfare (NEW), USA, 1965	53
2.9	Modification of GDP, Indonesia and Costa Rica	54
3.1	Selection and aggregation of social indicators	66
3.2	Human Freedom Index (HFI)	68
3.3	Activity analysis of feasible development	74
3.4	Frontier economics vs. deep ecology	77
4.1	Policy directions for sustainable development	80
4.2	Models of macroeconomic management for developing countries	86
4.3	Growth vs. environment	88
4.4	Environmental accounting and input–output analysis	89
4.5	Transfer pricing in PNG	92
4.6	Environmental exceptions to GATT obligations (Article XX)	93
4.7	Debt-for-nature swaps	96
4.8	Instruments of environmental cost internalization	102
4.9	Features and economic effects of a carbon tax in the USA	105
5.1	Ivory-tower plans: the reasons for plan failures	107
5.2	Selected goals for communicable disease control	114
5.3	UNCED programmes for sustainable human settlements development	115
5.4	Evolution, stability and exploitation of ecosystems	118
5.5	Integrated regional development in OAS member states	125
5.6	Environmental Impact Assessment (EIA)	128
5.7	Barriers to technology transfer	131
5.8	Environmental statistics and indicators	132
5.9	Green Plan, Canada – goals and selected initiatives	135
5.10	National Action Plan for Sustainable Development, Papua New Guinea – major themes and priority actions	136

5.11 The Declaration of Environmental Policy Regulation, Namea
     Landowners Association Incorporated   138
6.1 International funding mechanisms   144
6.2 UNCED conventions and principles   147
6.3 A Development Security Council   150
6.4 Closing statements at UNCED   151

FOREWORD

It is now over a year since the final gavel came down on the United Nations Conference on Environment and Development, that historic Earth Summit which fanned the embers of environmentalism into a blaze of worldwide enthusiasm and resolve to put a stop to humankind's plunder of the Earth.

Optimism was high as the Earth Summit concluded. But what of the follow-up?

At the political level – and particularly among national governments – movement towards implementation of the principles contained in the Declaration of Rio and the action programme incorporated in Agenda 21 has not been conspicuous. In the year since Rio we have witnessed a tendency on the part of individual governments to lapse back to business as usual – not surprising, though nevertheless disappointing, in light of the immediate political and economic problems which continue to plague all governments. Notably, the large-scale commitments of new financial resources required for implementation of Agenda 21 have not been forthcoming. Indeed, a number of countries – including my own – have cut back on their Official Development Assistance.

But there have also been some unmistakably positive signs. In June 1993 – on the first anniversary of the Earth Summit – the new administration in the United States signed the Treaty on Biodiversity, thus reversing that country's disappointing abstention at Rio. President Clinton has also formed the President's Council on Sustainable Development, a board including representatives of industry, government and environmental groups, charged with helping to formulate a national Sustainable Development Action Plan by the end of 1993.

At the international level, the United Nations has established its Sustainable Development Commission – an important recognition of the overriding importance of the subject. The commission is, of course, an intergovernmental body, but I hope that it will be open to the kinds of inputs by non-governmental actors that were the hallmark of Rio.

But if there is little encouragement to be drawn from the response of governments, and some caution regarding the efficacy of the UN

Commission, there is a great deal of promise in what is happening at the grass-roots level. In the longer term, there is real basis for hope that the Earth Summit has laid the foundations for the changes in public attitudes and the political mind-set necessary to achieve the transition to sustainable development.

Seminars, conferences and new initiatives in follow-up of Rio are proliferating throughout the world. Rio has produced a broad and growing global constituency committed to fulfilment of the hopes and expectations that the Earth Summit engendered.

The spirit of Rio burns on undiminished, now more than a year later. Indeed the Earth Summit has produced a universal dialogue of unheard-of intensity and determination. The protocols achieved at Rio, along with the Agenda 21 implementation programme, are now on everyone's agenda, at every level of society and in every corner of the Earth. The issue of sustainable development, which languished as an abstruse and idealistic notion for five years after it was propounded in the Brundtland Report in 1987, has now blossomed full-force as an inescapable and crucial imperative. As Emile Zola wrote nearly a century ago, 'Truth is on the march and nothing can stop it.'

It is clear that, while governments may be approaching sustainable development with understandable wariness, ordinary people, in communities and special interest organizations around the world, are not about to let the spirit of Rio die on the vine. They are translating it into a new era of grass-roots interest and action that will contribute to the implementation of Agenda 21, and infuse the political process with new energies and fresh resolve.

The Earth Summit made it clear that we have reached a crossroads in the human experience. Human activities have brought the world to this critical juncture, and human activities are now the principal determinant of whether the future of our planet will be a secure and hospitable home for humankind. We are literally in command of our own evolution.

While Agenda 21 has provided the bricks and mortar with which to build a more sustainable future, we do not yet have the detailed blueprints. These will emerge from the global dialogue which gains momentum with every passing day. An important part of this dialogue is the serious thought and penetrating analysis we are increasingly finding in such works as Peter Bartelmus' *Environment, Growth and Development: The concepts and strategies of sustainability*.

From its sobering assessment of the current state of the world's environment, through its examination of some of the thorny problems such as accounting for environmental costs, to its call for new global instrumentalities for co-operation, this book is a worthy and thought-provoking contribution to the vital debate.

Maurice F. Strong
August 1993

PREFACE

One year after the publication of *Environment and Development* in 1986, a radically different approach to dealing with environmental destruction was set in motion by the World Commission on Environment and Development, popularly known as the Brundtland Commission. Born out of frustration about continuing environmental decline and failures of international development strategies, the Commission proposed, under the label of 'sustainable development', to shift the focus from weak and peripheral environmental management to the socio-economic policy sources of environmental impacts. The idea of sustainable development is thus to integrate environmental concerns into mainstream policies. This can be seen as a reversal of the past approach to *environment* and development to one of *development* and environment.

As could be expected, such a move would not go unopposed by those 'environmentalists' who feared that the environmental cause would be smothered by economics. Typically, this discipline had looked upon the environment as an 'externality' only. A more balanced view seems to have prevailed, however, in the international discussion. The Earth Summit on Environment and Development in Rio de Janeiro stressed the need for striking a balance between environment and development by adopting sustainable development as its leitmotiv.

This book presents therefore a reappraisal of environment and development in terms of their sustainability. It calls for environmentally sound and sustainable economic growth and development in both developing and industrialized countries. This is a major expansion of the scope of *Environment and Development* which dealt mainly with the environmental problems of developing countries.

The meaning of sustainable development and its implications for policy-making are not yet clearly understood. This has not discouraged the general acceptance of the concept by all kinds of groups and organizations. For instance, 'industrialists' considered sustainable development as providing investment opportunities in a new industry of environmental protection; on the other hand, 'environmentalists' saw sustainable development as a new paradigm of development that would replace economic growth by 'qualitative' development, allowing human beings to live in greater harmony with nature. This book sheds some light on the concepts of sustainable development and sustainable economic growth by means of more rigorous definitions and use of operational variables in integrated economic–environmental analysis. To a great extent, this is facilitated by new 'green accounting' methodologies in whose development I had the privilege to be actively involved.

Environment and Development was originally conceived out of frustration about more than one decade of rhetorics (since the first global 'Stockholm Conference' on the human environment in 1972) about the need for integrated development. The answer then found was to propose a shift from ineffective central planning and policies to focusing on local-level management of natural resources and economic activity. 'Ecodevelopment' was considered a practical way of directly improving conditions of poverty *in situ* and in sympathy with custom, tradition and environment. It is maintained that none of the arguments for local-level ecodevelopment have lost their validity. The limited 'reach' of monetary accounting and economic instruments of environmental cost 'internalization' make grass-roots ecodevelopment a necessary complement to central planning and policies of sustainable growth and development.

In order to link systematically the different policies, strategies and programmes of growth and development an integrative policy framework for sustainable development at national, sub-national and international levels is advanced. The framework's focus on operational, quantifiable concepts and methods should facilitate a rational assessment of such development. The framework should reveal misunderstandings, biased views and contradictory elements of this complex new paradigm. It would indeed be a tragedy if, at a time when East–West conflicts have dramatically decreased, new tensions between North and South would prevail on such issues as responsibility for global environmental problems and environmental 'conditionalities' in development assistance.

The first chapter examines the evidence for the perceived non-sustainability of human activity on Earth. It finds that there is no single indicator, but rather an abundance of sometimes contradictory data. Such data alert to disturbing trends in natural resource depletion and environmental degradation but do not permit an overall evaluation of the sustainability of economic growth or development. Chapter 2 explores, therefore, possibilities of developing measures that compare the 'goods' of economic production and consumption with the 'bads' or costs of losses of natural resources and environmental quality. The new System of integrated Environmental and Economic Accounting (SEEA) is described and critically reviewed in this regard.

One result of such integrated accounting are definitions of Environmentally-adjusted Domestic Product and National Income indicators. Those indicators provide the basis for definitions of sustainable economic growth in Chapter 3. Limitations of monetary valuation of non-marketed environmental effects require the definition of sustainable 'development' in non-monetary terms. Such definition shifts the analysis from sustainability to the feasibility of development programmes complying with environmental and socio-economic standards.

Chapters 4 and 5 examine policies that suggest themselves as a result of the conceptual analysis of Chapter 3. A distinction is made, within an overall policy framework, between policies of sustainable economic growth (Chapter 4) and sustainable development (Chapter 5). The role of environmentally-adjusted macroeconomic variables in policies of growth, employment, trade and foreign debt, and the use of market instruments for environmental cost internalization and structural change are discussed in Chapter 4. The limited reach of economic policies is the reason for elaborating strategies and action plans of integrated development and local-level ecodevelopment in Chapter 5. The last chapter addresses the global environmental concerns raised in the first chapter. It briefly reviews the implications of the Rio Summit for national and international policies and proposes a global compact among nations for the implementation and monitoring of the Rio recommendations.

Note

The views expressed here are the author's and do not necessarily reflect an expression of opinion on the part of the United Nations.

ACKNOWLEDGEMENTS

The author and publisher would like to thank the following individuals and organizations that have given permission for the reproduction of copyright material:

Smith/Gibbs publishers (Box 3.4); Organization for Economic Co-operation and Development (Table 4.3); United Nations Educational, Scientific and Cultural Organization (UNESCO) (Fig. 1.1); United Nations Environment Programme (UNEP) (Fig. 1.4); World Bank (Table 2.3); World Resources Institute (Figs 1.3 and 1.5).

ABBREVIATIONS

EDP	Environmentally-adjusted (net) Domestic Product
EIA	Environmental Impact Assessment
ENI	Environmentally-adjusted National Income
FAO	Food and Agriculture Organization of the United Nations
FDES	Framework for the Development of Environment Statistics
FRG	Federal Republic of Germany
GATT	General Agreement on Tariffs and Trade
GDP	Gross Domestic Product
GEF	Global Environment Facility
GNP	Gross National Product
HDI	Human Development Index
HFI	Human Freedom Index
IIASA	International Institute for Applied Systems Analysis
ILO	International Labour Organization
IPCC	Inter-governmental Panel on Climate Change
ISEW	Index of Sustainable Economic Welfare
LDC	Least Developed Country
NDP	Net Domestic Product
NEW	Net Economic Welfare
NGO	Non-Governmental Organization
NI	National Income
NIEO	New International Economic Order
NIMBY	Not-In-My-Backyard
NPP	Net Primary Production
NSDS	National Sustainable Development Strategy
OAS	Organization of American States
OECD	Organization for Economic Co-operation and Development
OECS	Organization of Eastern Caribbean States
PNG	Papua New Guinea
PRED	Population, Resources, Environment and Development
SAM	Social Accounting Matrix
SEEA	System of integrated Environmental and Economic Accounting

SNA	System of National Accounts
SNI	Sustainable National Income
UNCED	United Nations Conference on Environment and Development
UNCHS	United Nations Centre for Human Settlements (Habitat)
UNCTAD	United Nations Conference on Trade and Development
UNCTC	United Nations Centre on Transnational Corporations
UNDP	United Nations Development Programme
UNEP	United Nations Environment Programme
UNESCO	United Nations Educational, Scientific and Cultural Organization
UNFPA	United Nations Population Fund
UNSTAT	United Nations Statistical Division
WCED	World Commission on Environment and Development
WRI	World Resources Institute

1

WHAT ON EARTH IS WRONG?

This chapter describes the main concerns with development, growth and the environment. An overview of international strategies and approaches reveals large-scale failures in all those areas. Selected indicators of the state and trend of the environment try to answer the questions of what and how much is wrong with the planet. The chapter concludes that the latter (how much?) requires integrated databases and aggregated indicators as common yardsticks of environment and development.

1.1 Development and environment: from global discussion to global frustration

Lost decades of development

Economists and parliamentarians agree that the 'past decade has been a cruel disappointment' to developing countries (*The Economist*, 23 September 1989) – indeed 'a lost decade of development' (Parliamentarians for Global Action 1990). With a few exceptions of developing countries in South-east Asia, the rich countries got richer and the poor ones poorer.

Before attempting to determine who are the winners and losers in the global strife for growth and development, it is useful to recall what is commonly understood by economic growth and development. Development is generally accepted to be a process that improves the living conditions of people. Most also agree that the improvement of living conditions relates to non-material wants as well as to physical requirements. Development goals that call for the increase of human welfare or the improvement of the quality of life reflect this agreement.

Quantifiable definitions of these concepts are needed for measuring progress towards development goals. A typical starting point has been to break down the overall objective of human welfare into sub-objectives or targets. The difficulties involved are described by Bartelmus (1980: 40 *et seq.*). Box 1.1 offers a tentative list of general (primary) human objectives, condensed from a variety of publicly proclaimed social objectives. Subjective

---

Box 1.1 Basic human objectives

Primary objectives

- Affection/love.
- Recreation/entertainment.
- Education.
- Human freedoms (security).
- Shelter.
- Aesthetic/cultural values.
- Political equity (participation, social opportunities).
- Health.
- Physiological needs.
- Future quality of life.

Secondary objectives can be considered as instrumental in fulfilling primary ones. They include food, clothing, transportation, drinking-water, social services, housing, conservation of the environment, maintenance of stability and justice, nation building, distribution of income and wealth, social security, working conditions, employment, time and leisure, education and training, and security.

Source: Bartelmus (1980: 46).

---

value judgements are involved in such a list. Any further breakdown would be even more arbitrary as human preferences for more specific 'secondary' objectives or desirables (as enumerated in the box) vary significantly among individuals and through time and space.

Generally, applicable policies and strategies to meet such objectives are as difficult to identify as the objectives themselves. Such policies must weigh trade-offs between goals and values within changing socio-economic conditions. In most developing nations, low levels of living and productivity are accompanied by high levels of population growth, unemployment, international dependence and a predominantly agrarian base to the economy. Based on these common factors, some international agreements on development strategies have been reached. However, those agreements had to be revised repeatedly in view of considerable failures of the proposed strategies at the national level.

As indicated in Box 1.2, this stark picture of development seems to have brought back a focus on economic growth, the central theme of the First International Development Strategy in the 1960s. The Fourth Strategy for the 1990s advocates the revitalization of economic growth while paying some attention to other 'aspects' of development. To some extent, this return to economic growth strategies might have been brought about by the absence of a widely accepted indicator of development. The

---

Box 1.2 International Development Strategies of the United Nations

The International Development Strategy of the *First United Nations Development Decade* of the 1960s was based on the belief that the fruits of accelerated economic growth would trickle down to the low-income population strata. Since the trickle-down effect did not materialize and thus failed to alleviate poverty in developing countries, the objective of social justice was included in the objectives of the *Second Development Decade*. The idea was to improve the distribution of the results of economic growth. The strategy for the *Third Development Decade* of the 1980s proceeded from the recognition that the objectives of the Second Decade were not met because of inequities and imbalances in international economic relations. The strategy, therefore, included the goal of establishing a 'New International Economic Order'. However, the New Order also failed to materialize, and the latest attempt at an International Development Strategy for the 1990s (the *Fourth United Nations Development Decade*) had to admit that 'the goals and objectives for the Third United Nations Development Decade were for the most part unattained' (United Nations, General Assembly resolution 45/199). The 1980s were characterized as a decade of falling growth rates, declining living standards and deepening poverty with a widening gap between rich and poor countries. Accordingly, the latest strategy for the 1990s calls again for the acceleration of economic growth. Such growth is also seen as a prerequisite for 'priority aspects of development' that include the eradication of poverty and hunger, human resources development and the protection of the environment.

---

economic-growth concept of development utilizes Gross Domestic or National Product as a concise measure, usually in per-capita and real (price-deflated) terms. It is generally acknowledged, however, that economic growth is at best an 'essential' (World Bank 1992: 34) or a 'mere' (UNDP 1992b: 2) means of development rather than an end in itself.

For a multidimensional concept of development, as expressed in lists of human objectives or needs, it is more difficult to find a similar aggregate measure. As shown in Chapters 2 and 3, aggregate development indicators and the correction of national accounting aggregates to obtain measures of economic welfare are still in their infancy; they cannot (yet) replace national or domestic product, which provide an overall insight into a country's productive capabilities and hence into one of its major sources of national welfare. Nonetheless, some attempts at categorizing countries in terms of overall 'development indices' have been made and can be compared to rankings of countries in terms of conventional domestic product or national income.

Three approaches to classifying countries according to their stage of development are compared in Table 1.1. The basic grouping is determined by the traditional measure of Gross Domestic Product (GDP) per capita at purchasing power parities rather than conventional exchange rates (United

Table 1.1 Country categories by level of growth and development – GDP per capita[a] (A to E), HDI[b] (a to e) and LDCs[c](*)

A $244–749	B $773–1777	C $2011–3476	D $3619–9760	E $9848 +
*Afghanistan a	*Bangladesh a	Algeria c	Argentina d	Australia e
Angola a	*Benin a	*Botswana c	Barbados e	Austria e
*Burkina Faso a	Bolivia b	Dominican	Bahrain d	Belgium e
*Burundi a	Cameroon b	Rep. c	Brazil d	Canada e
*Cent. Afr.	Congo b	Ecuador c	Chile d	Denmark e
Rep. a	Cote d'Ivoire b	Gabon c	Colombia d	Finland e
*Chad a	Egypt b	Guatemala c	Costa Rica d	France e
*Ethiopia a	El Salvador c	Indonesia c	Cyprus e	Federal
Ghana b	*Gambia a	Iraq c	Fiji d	Republic of
*Guinea a	Guyana c	Jamaica d	Greece e	Germany e
*Madagascar b	*Haiti b	Jordan c	Iran c	Iceland e
*Malawi a	Honduras c	Mauritius d	Ireland e	Israel e
*Mali a	India b	Morocco b	Malaysia d	Italy e
*Mauritania a	Kenya b	Nicaragua c	Mexico d	Japan e
*Mozambique a	*Lesotho b	Panama d	Portugal d	Kuwait d
*Myanmar b	*Liberia b	Papua New	Saudi Arabia d	Luxembourg e
*Niger a	*Nepal a	Guinea b	South Africa d	Malta d
Nigeria b	Pakistan b	Paraguay c	Syrian Arab	Netherlands e
*Rwanda a	Senegal a	Peru c	Rep. c	New Zealand e
*Sierra	*Togo b	Philipines c	Suriname d	Norway e
Leone a	*Uganda a	Sri Lanka c	Trinidad and	Oman c
*Somalia a	*Zambia b	Swaziland c	Tobago d	Singapore d
*Sudan a	Zimbabwe b	Thailand d	Turkey c	Spain e
*Tanzania (UR) b	*Cape Verde b	Tunisia c	Uruguay e	Sweden e
*Zaire b	*Comoros b	*Yemen b	Venezuela d	United
*Bhutan a	*Laos PDR b	*Maldives c		Kingdom e
*Cambodia a	*Sao Tome and	*Samoa c		United States e
*Djibuti a	Pr. b	*Vanuatu c		
*Equ. Guinea a	*Solomon			
*Guinea-Bissau a	Islands b			
*Kiribati				
*Tuvalu				

Notes
[a] 1989, at purchasing power paraties (*source*: United Nations, 1993b: table A.1); countries for which GDP per capita at purchasing power parities was not available are listed in italics.
[b] Human Development Index (*source*: UNDP 1992b).
[c] Least Developed Countries as established by the United Nations General Assembly (*source*: United Nations 1992d).

Nations 1993b). Countries are grouped into five categories from A for the poorest and E for the richest countries. This ranking by economic growth is contrasted with a broader development measure, the Human Development Index (HDI), developed by the United Nations Development Programme (UNDP 1992b). The Index combines indicators of national income, life expectancy and educational attainment. It reflects, therefore, a relatively arbitrary selection of factors considered to contribute 'significantly' to development. A more politically determined categorization of Least Developed Countries (LDCs) is established by the General Assembly of the United Nations. LDCs are countries with low GDP per capita, a low level of the 'physical quality' of life (life expectancy, calorie supply, school enrolment and adult literacy) and lack of economic diversification. Those countries are indicated with an asterisk in Table 1.1.

LDCs are closely correlated with the level of economic growth (categories of A and lower ranking B). Notable exceptions are a few small island developing countries (Maldives, Samoa, Vanuatu) where the low diversification aspect (and dependence on development assistance) (United Nations 1992d: 61) has prevailed in the selection. In terms of the broad categories (A to E and a to e), there seems also to be a close relation between economic growth and human development. One major exception is the muted significance of high-level income in oil-rich Middle East countries in the HDI: those countries lose up to 37 ranks (in the original tables) due to the reduced significance of high incomes in the HDI.

Environmental doomsday and international reaction

It is worthwhile to recall the ups and downs of the environmental movement because there seems to be some risk of recurrence of early (over)reaction to environmental problems. Conspicuous pollution incidents in the 1960s and neo-Malthusian responses to demographic and economic growth led to the appearance of environmental doomsday literature. Titles like *The Death of Tomorrow* (Loraine 1972), *Silent Spring* (Carson 1965) or *Blueprint for Survival* (Goldsmith *et al.* 1972) are indicative of the environmental mood in the late 1960s and early 1970s. The use of a seemingly objective computerized global model was probably responsible for provoking much of the widespread attention to the Club of Rome's *Limits to Growth*, (Meadows *et al.* 1972). The model predicted a 'rather sudden and uncontrollable decline in both population and industrial capacity' within the next century if current growth trends remain unchanged. Zero (or even negative) growth was advocated to avoid the disastrous consequences of transgressing the physical limits of the Earth's resources.

Another pessimistic conservationist view focused on the preservation of ecosystems and endangered species to the neglect of socio-economic conditions and consequences. Examples of this view are Curry-Lindahl's *Conservation for Survival* (1972) and Caldwell (1972) who purports to defend Earth against the 'unecological animal' man.

Those policies could not be accepted by countries that were still in the early stages of socio-economic development. For them, economic growth appeared to be more important than concern about a few endangered species of wildlife. Only affluent countries were seen to be able to afford the luxury of diverting some of their wealth to environmental protection. Moreover, the high and wasteful consumption levels of the industrialized nations placed a large stress on the resources of developing countries. Proclamations of global solidarity for spaceship Earth were thus met with suspicion and distrust by developing countries. The only view rich and poor countries seemed to share was the conviction that environmental conservation and economic development were in conflict with each other (UNEP 1978). It is the merit of the international community that it has opened a dialogue on the environment-and-development issue between developed and developing countries through a number of international seminars and conferences.

The Secretariat of the United Nations Conference on the Human Environment organized a seminar on development and environment at Founex, Switzerland (United Nations Conference on the Human Environment 1972). The seminar concluded that environmental problems do not only result from the development process itself but also from the very lack of development. Poor water, inadequate housing and sanitation, malnutrition, disease and natural disasters were cited. The term 'pollution of poverty' was later coined to describe this aspect of the environmental question. Consequently, environmental goals would provide a new dimension to the development concept itself, requiring an integrated approach to environment and development. The United Nations Conference on the Human Environment (United Nations 1973) endorsed these principles, emphasizing that environmental and developmental goals could be harmonized, *inter alia*, by the wise use of natural resources. The Conference also established a small (but rapidly expanding) secretariat, the United Nations Environment Programme (UNEP) to facilitate and monitor the implementation of the Conference's action plan.

In the wake of the energy crises in 1973 and 1974 and of the declaration of a New International Economic Order in 1974 (see Box 1.3), a joint UNEP/UNCTAD Symposium at Cocoyoc attempted to integrate a general assessment of development goals with new ideas in the field of environment. It was recognized that the failure of society to provide a safe and happy life for all is not one of 'absolute physical shortage but of economic and social maldistribution and misuse'. Hence, the Symposium advocated a strategy of satisfying first basic human needs, with due consideration for global environmental risks or so-called 'outer limits' (UNEP and UNCTAD 1974). The basic-needs approach was taken up and widely publicized by the Programme of Action of the 1976 World Employment Conference. The Programme recognized food, shelter, clothing and essential services such as safe drinking-water, sanitation, transport, health and education as basic human needs and requested that basic-needs policies become an essential part of the United Nations Third Development Decade Strategy (ILO 1977).

Since then, international statements have tended to dissociate themselves from the basic-needs approach. Developing countries considered the strong support of this strategy by developed nations as a tactical move of diverting

---

Box 1.3   New International Economic Order (NIEO)

As early as 1974, the General Assembly of the United Nations adopted a declaration, on the NIEO (General Assembly resolution 3201, S–VI). The reason was that 'it has proved impossible to achieve an even and balanced development under the existing international economic order'. This order was characterized by inequality, domination, dependence, narrow self-interest and segmentation (United Nations 1982). Consequently, the NIEO called for equity, sovereign equality, interdependence, common interest and co-operation among all states. Key principles on which such co-operation should be based included, *inter alia*,

- full permanent sovereignty of every state over its natural resources;
- regulation and supervision of transnational corporations;
- just and equitable prices of raw materials and other export commodities;
- access to the achievements of science and technology.

---

attention away from the implementation of the New International Economic Order (NIEO) (see Box 1.3). In addition, they felt that an international strategy concentrating only on the satisfaction of basic needs represents an intrusion into sovereign national development policies. Thus the International Development Strategy for the Third Development Decade omitted any reference to basic human needs, stressing instead the establishment of an NIEO. However, as indicated in Box 1.2, negotiations on the implementation of the NIEO did not materialize, and the Fourth Development Strategy simply referred the discussion of environment and development to the United Nations Conference on Environment and Development in 1992.

Search for a new paradigm: sustainable development

Despite the calls for the integration of environmental and developmental objectives, the fact remains that in general such integration did not take place. Issues of population growth and concentration, desertification, pollution and resource exploitation continued to be the responsibility of specialized departments, while macroeconomic policies focused on the maximization of economic growth. Environmental impacts were addressed to some extent by environmental agencies without much influence, however, on the process of socio-economic decision-making in central government. 'A widespread feeling of frustration and inadequacy in the international community about our own ability to address the vital global issues and deal effectively with them' (WCED 1987: ix) had therefore been the basic motivation to establish in 1983 a World Commission on Environment and Development (WCED) of the United Nations.

Under the generic label of sustainable development, the WCED proposed

7

Box 1.4   Objectives of sustainable development

Critical objectices for environment and development policies that follow from
the concept of sustainable development include:

- reviving growth;
- changing the quality of growth;
- meeting essential needs for jobs, food, energy, water and sanitation;
- conserving and enhancing the resource base;
- reorienting technology and managing risk; and
- merging environment and economics in decision-making.

Source: WCED (1987: 49).

---

a large variety of policy recommendations that would sustain the environmental base of development. 'Critical objectives' for sustainable development, proposed by the WCED, are listed in Box 1.4. As indicated above,
the idea of effectively merging environmental issues into socio-economic
planning and policies is not new. What is new is the approach advocated
for implementing sustainable development. This approach calls for a move
from dealing with environmental effects, after their occurrence, to focusing
on the 'policy sources' of these effects for preventive action (*ibid.*: 310 *et seq.*).
This can also be seen as a shift from a discussion of *environment* and development to *development* and environment. In other words, this is an attempt to
merge environmental issues with mainstream policy-making rather than to
change those policies from the periphery of the environmental movement.

The recent United Nations Conference on Environment and Development (UNCED), the Earth Summit in Rio de Janeiro, attempted to translate
the new paradigm of sustainable development into a globally adopted
philosophy, an 'Earth Charter' and an international action programme, the
'Agenda 21'. The final chapter of this book examines to what extent these
ambitious goals were achieved in Rio. Judging from the flurry of publications on the sustainability or non-sustainability of economic growth and
development, it appears that there is no consensus yet on sustaining current
socio-economic activities and trends, nor on the exact meaning and implications of the new concept of sustainable development. The following section
attempts to shed some light on the former by analysing the current state and
possible unsustainable trends of environment and development.

1.2 Diagnosis: a planet at risk

Reaching the limits?

Since the publication of the report of the WCED (1987), the environment
has been in the limelight of public attention. A good example is *Time*

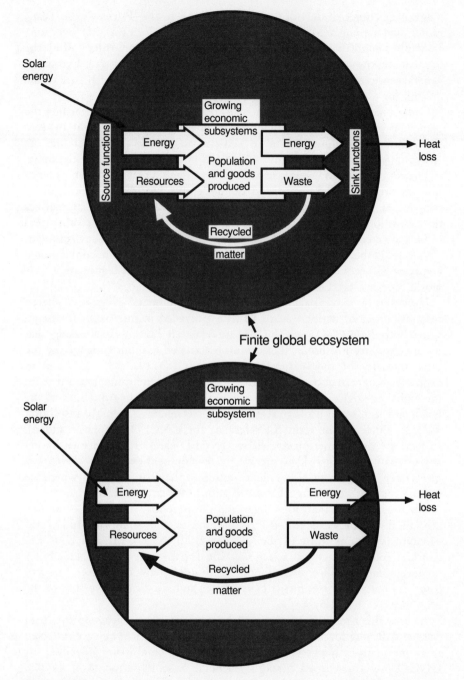

Figure 1.1 The finite global ecosystem relative to the growing economic subsystem
Source: Goodland (1991: 17).

magazine's choice of devoting an entire edition to the 'Planet of the Year', rather than naming a 'Man of the Year'. The reason given was that 'everyone suddenly sensed that this gyrating globe, this precious repository of all the life that we know of, was in danger' (*Time*, 2 January 1989: 27). Evoking a general sense of danger and calls for 'a war for survival' (*ibid*.: 30) are fertile ground for a return to the doomsday scenarios of the 1960s.

Other voices joined more recently in sounding the red alert regarding the planet's capacity to deal with environmental stress. Brown (1993: 16) sees the existing economic system in self-destruction 'as it undermines its environmental support systems'. Goodland (1991: 5) insists that 'the limits to growth have already been reached'. In his view, there is plenty of evidence that we have made the transition from an 'empty world' to a world full of people that stress the biosphere's sources and sinks. Figure 1.1 dramatizes this transition as an increase in the relative size of the economic subsystem of the global ecosystem, leaving hardly any space for a pristine environment. This view is shared by Daly who anticipates 'the death of the conventional empty-world economists to be replaced by the – not yet legitimate – full-world economists' (Daly 1991b: 30).

However, a more balanced view of environmental risks and shared responsibilities of nations seems to have prevailed in international discussions. Both the WCED and UNCED are relatively cautious in stressing 'that we are close' to thresholds 'that cannot be crossed without endangering the basic integrity of the [natural] system' (WCED 1987: 32, 33), and in expressing concern about 'trends that, if allowed to continue, *could* disrupt the global ecological balance, jeopardize the life-sustaining qualities of the Earth and lead to an ecological catastrophe' (General Assembly resolution 44/228, which set the terms of reference for UNCED; emphasis added). In fact, the documents produced by UNCED carefully avoid mentioning any ultimate limits or risks to growth or development except for a relatively meek reference to 'continuing deterioration of the ecosystems on which we depend for our well-being' (United Nations 1993c: para. 1.1).

Obviously, the conclusions regarding the potential for future economic growth differ between full-world economists and somewhat-full-world economists such as those of the WCED. Both the WCED and UNCED consider economic growth a *sine qua non* for the long-term sustainability of development. Full-world economists disagree, advocating zero growth (but 'qualitative' development: Daly 1991c) and a radical conversion of the economy (see Box 1.5).

At least this time, there appears to be worldwide consensus on the main risks and driving forces of non-sustainable development. They are reflected in an internationally agreed list of nine environmental issues, specified for UNCED by the United Nations (General Assembly resolution 44/228, para. 12). The issues (listed in Box 1.6) are considered to be the main concerns in achieving environmentally sound and sustainable development.

Box 1.5   Facing up to change

We know what we have to do. And we know how to do it. If we fail
to convert our self-destructing economy into one that is environmentally sustainable, future generations will be overwhelmed by environmental degradation and social disintegration. Simply stated, if our
generation does not turn things around, our children may not have the
option of doing so.

(Brown 1993: 21)

Box 1.6   Environmental issues (slightly abbreviated)

1 Protection of the atmosphere by combating climate change, depletion of
   the ozone layer and transboundary air pollution.
2 Protection of the quality and supply of freshwater resources.
3 Protection of the oceans and all kinds of seas, including coastal areas, and
   the protection, rational use and development of their living resources.
4 Protection and management of land resources by, *inter alia*, combating
   deforestation, desertification and drought.
5 Conservation of biological diversity.
6 Environmentally sound management of biotechnology.
7 Environmentally sound management of wastes (hazardous and toxic), as
   well as prevention of illegal international traffic in toxic and dangerous
   products and wastes.
8 Improvement of the living and working environment of the poor in urban
   slums and rural areas, through the eradication of poverty.
9 Protection of human health conditions and the improvement of the quality
   of life.

Source: United Nations, General Assembly resolution 44/228.

The list has been criticized for omitting social and demographic concerns,
energy production and consumption and natural disasters. Some of these
questions, such as population and poverty and their interaction with
environmental issues, were taken up later by UNCED. Others (energy or
natural disasters) were dealt with less explicitly under other pertinent topics,
for example, climate change or the protection of land resources.

Both poverty and affluence were identified as the driving forces behind
environmental degradation and resource depletion, sweepingly termed
pollution of poverty and pollution of affluence. The former refers to
pressures of growing populations in poor countries on marginal and
vulnerable lands, forests and congested cities. Thus many environmental
problems in developing countries originate from the struggle to overcome
extreme conditions of poverty. In industrialized countries, on the other

11

hand, impacts of high-level economic growth and consumption are responsible in most cases for environmental deterioration. There is also a difference in the significance of environmental impacts: wealthy countries face a deterioration of the quality of life, but life itself may be at stake in developing countries when their natural resource base is destroyed.

A number of international surveys, seminars and studies have been undertaken for the worldwide assessment of the state and trends of the environment. The results of these surveys and studies suggest that most environmental impacts can be observed in the developing as well as the developed regions of the world. Environmental problems differ, however, in scope and intensity. Consequently, they receive differing priorities in national planning and policy-making. Table 1.2 summarizes the issues discussed in those assessments, pointing out (by capital letters) concerns that are considered to be of particular significance. The depletion and degradation of natural resources (land/soil, water and forests) and their effects on food and energy supply, marginal conditions in human settlements, environmentally conditioned diseases and natural disasters are high-priority issues in developing countries. By contrast, industrialized countries are especially concerned about air, land and water pollution, global environmental phenomena of climate change and the depletion of the ozone layer.

Industrialized countries have been blamed for the global impacts of pollution by developing countries. The term 'common but differentiated responsibilities' in the Rio Declaration of UNCED (United Nations 1993c) refers to this accusation. 'Denial' of these concerns and responsibilities by the world's political and business leaders has been considered 'among the most paralyzing of human responses' (Postel 1992: 4). It appears that the UNCED has indeed overcome such denial by acknowledging the problem and its root causes (see below, Chapter 6). What is needed now is to replace dramatization (as those of Fig. 1.1) and rhetorics (as those of Box 1.5) with scientific measurements, synthesized into statistics and indicators that provide a comprehensive picture of the sustainability or non-sustainability of current human activities. The limits of obtaining such a picture with existing data are discussed in the following.

Indicators of global (non-)sustainability

Most of the evidence given for reaching global limits of economic growth refers to exhaustible energy resources, global warming from climate change, depletion of the ozone layer, land degradation and species extinction. The above-described full-world hypothesis is largely based on indicators of the human appropriation of biomass produced by photosynthesis. The following examines to what extent this evidence is supported or disputed by available environmental indicators and their interpretation. Table 1.3 provides quantitative information that is typically offered as evidence for the non-sustainability of growth or development.

Table 1.2 Environmental concerns of developing and industrialized countries

Environmental concerns	Developing countries	Industrialized countries
I Natural environment		
A Air	Air pollution in major cities	AIR POLLUTION
B Land, soil, mineral resources (incl. energy)	SOIL EROSION AND DEGRADATION; DESERTIFICATION	Soil loss and deterioration; dumping of waste; risk of radioactive contamination from nuclear-power production
C Water	FRESHWATER SHORTAGE; freshwater pollution (sewage, pesticides); pollution of coastal waters	Freshwater shortage; INLAND AND MARINE WATER POLLUTION
D Fauna and flora	DEFORESTATION (especially of tropical forests); loss of genetic resources; endangered species	Loss of genetic resources; endangered species
E Ecosystems	Pollution of coastal ecosystems (decreasing fish catch)	Disruption of mountain, wetland, freshwater (especially FOREST DAMAGE from acid rains and eutrophication) and coastal ecosystems
F Natural disasters	FLOODS; DROUGHTS; STORMS; EARTHQUAKES; volcanic eruptions	Floods; earthquakes
II Man-made environment and living conditions		
A Bioproductive systems	LOSS AND DEGRADATION OF ARABLE LAND; pests and pest resistance; water shortage, pressures on fish population (over-fishing, pollution); IMPACTS OF FUELWOOD CONSUMPTION; food contamination, post-harvest losses	Loss of croplands to urban sprawl; pests and pest resistance; contamination of crops and fish; over-exploitation of fishing grounds
B Human settlements	MARGINAL SETTLEMENTS (RURAL–URBAN MIGRATION, URBAN GROWTH)	URBAN SPRAWL; NOISE; LAND CONTAMINATION; TRAFFIC CONGESTION
C Health	MAL– AND UNDERNUTRITION; INFECTIOUS AND PARASITIC DISEASES	CANCER; cardiovascular diseases; genetic and long-term effects of POTENTIALLY TOXIC CHEMICALS and HAZARDOUS WASTE
D Environment and development	SUSTAINABLE DEVELOPMENT	ENVIRONMENTAL EXTERNALITIES; energy and environment
III Global problems	Global warming and consequential effects	CLIMATE CHANGE; depletion of the OZONE LAYER

Sources: Bartelmus (1986: table 2.1); OECD (1992); UNEP (1992).

Table 1.3 Selected indicators of (non-)sustainability

Environmental concern	Environmental impacts	Effects and repercussions	Evaluation (limits/thresholds)
BIOMASS APPROPRIATION	*Global annual net primary production used:* – directly (people and domestic animals): 3% – total appropriations: 40% (Vitousek et al. 1986)	*Extinction of species* *Genetic impoverishment* (see also below)	*Limits to growth not so distant (Vitousek et al. 1986: 373) *Limits:* 40% of biomass appropriation reflect a relatively full world; 80% represents excessive fullness (Daly 1991b: 18)
CLIMATE CHANGE	*Greenhouse gases,* increase in atmospheric concentration (1765–1990): – CO_2: 25% – CH_4: 100% – NO_x: 8% *Global warming,* mean global temperature increase: – past 100 years: 0.3–0.6°C – by year 2100: 2–5°C (best estimate 3°C)	*Sea-level,* mean global *increase* (business-as-usual): – by year 2030: 20 cm – by year 2100: 65 cm *Other:* decline and increase in agriculture, fauna and flora, water resources	*Business-as-usual scenario: possible catastrophic consequences of sudden adaptation of socio-economic structures (UNEP 1992: 22) *Global warming: evidence that limits to growth have been reached (Goodland 1991) *Costly greenhouse effects cannot be ruled out, nor compellingly demonstrated (World Bank 1992: 62) *Limits:* – global warming: 2°C (National Institute 1992: 146) – return to 1990 levels of greenhouse gas emissions by the year 2000 (envisaged protocols to Framework Convention on Climate Change)
OZONE LAYER	*Chlorine equivalent* concentration in the atmosphere: over 3.5 ppbv (National Institute 1992: 131)	*UV-B radiation:* – 1% reduction in stratospheric ozone leads to about 2% increase in UV-B reaching the ground	*Cosmic holes in the ozone shield are compelling evidence that human activity has already damaged our life support systems (Goodland 1991: 10)

	Ozone hole: average decrease of 30–40% in the total ozone column (15–20 km above Antarctica) Ozone decrease: – 5–10% in the upper atmosphere over much of the middle and high latitudes – average global depletion over the past 12 years: 3%	– exposure to UV-B can suppress the immune system, lead to cataracts and skin cancer, damage plants and aquatic organisms and reinforce the greenhouse effect	*The threat from UV radiation is certain to worsen, though longer-term changes associated with ozone depletion are difficult to detect (for example, due to ozone pollution) (World Bank 1992: 63) Limits: – maximum of 2 ppbv Cl equivalents to prevent permanent reduction of the ozone layer (National Institute 1992: 131) – phase-out of most depleting substances by the year 2000 (Montreal Protocol of the Vienna Convention for the Protection of the Ozone Layer)
LAND (DEGRADATION AND DESERTIFICATION)	Desertification: 5 million ha of irrigated and rainfed cropland lost annually Desertification and degradation: 70% of agriculturally used dryland affected Land degradation: 11% of the Earth's vegetated surface moderately or severely degraded since 1945 (WRI 1992: 3); 500 billion tonnes of top soil lost since 1972 (Brown 1993:4)	Environmental refugees: land degradation remains the single most important cause of about 10 million persons misplaced (Jacobson 1988)	*More widespread than desertification is the gradual deterioration of agricultural soils (World Bank 1992: 55) *There is no simple threshold of land sufficiency for food needs of future generations (WRI 1992: 3) Limits: population supporting capacity of lands in developing countries (excl. East Asia): – one and a half times of their year 2000 population (unrestricted movement of potential surplus food production) – 64 countries out of 117 have insufficient land resources to meet the food needs of their population by year 2000 (movement of potential surplus food within countries only) (FAO, UNFPA and IIASA 1982)

Table 1.3 Continued

Environmental concern	Environmental impacts	Effects and repercussions	Evaluation (limits/thresholds)
BIODIVERSITY (AND DEFORESTATION)	*Biodiversity*: perhaps a quarter of total biodiversity is in serious danger of extinction during the next 20–30 years *Species*: 5000 to 150,000 lost annually (Goodland 1991) *Deforestation*: annual deforestation rate (1981–90) of 16.8 million ha p.a.	*Contribution to development*: by wild species and their genetic variations in agriculture, medicine and industry *Other*: adaptation to unforeseen environmental stress, heritage for future generations, aesthetic rewards (World Bank 1992)	*Loss of biodiversity will compromise the ability of future generations to meet their needs (UNEP 1992: 55) *Biodiversity is complex beyond understanding and valuable beyond measure (Ryan 1992: 9) *Mass extinction has already begun (*ibid.*) *Limits*: sustainable maximum wood consumption of 6 billion m^3 (1988: total world consumption of roundwood of about 3 billion m^3) (UNEP 1992: 48)
ENERGY	*Reserves*, world (solids, oil, gas) at 1990 levels of consumption: proved recoverable: 90 years; proved in place: 243 years; total, including estimated additional: 800 years (United Nations 1992c) *CO$_2$ emission*: 9.1 billion tonnes of C in year 2005, may double in year 2010 (at current trends of energy use and efficiency) *Nuclear waste*: volume of high-level waste and spent fuel (in 1990) 21,000 m^3, could reach some 1 million m^3 (in 2000)	*Pollution*: urban air quality, acid deposits, climate change *Nuclear power*: low-level radiation effects on humans, safety of installations (including decommissioning)	*Use of fossil fuels is the best single example of the world being run unsustainably: oil and gas provide 60% of energy with barely 50 years of proven reserves (Goodland 1991: 5) *The world is not running out of marketed non-renewable energy and raw materials (World Bank 1992: 37) *Key limit to economic growth: sink constraint of fossil energy use (Goodland 1991: 6) *Uncertainty about the effects of exposure to low-level radiation (UNEP 1992: 111–12) *No disposal of high-level nuclear waste has taken place yet (UNEP 1992: 112)

Source: Where not otherwise indicated UNEP (1992).
Note: Billion equals 1000 million.

It has been argued that a doubling of the population within the next 35 years will exhaust the photosynthetic capacities for human activity (Goodland 1991; Daly 1991b). This stark prediction is derived from indicators of the human appropriation of biomass ranging between 3 per cent of the biosphere's annual net primary production (NPP), for the direct use of organic material by people and domestic animals, and 40 per cent of NPP if the 'co-option, diversion and destruction' of the total terrestrial resources is considered (see Table 1.3).

The latter figure appears indeed to be strong argument for full-world economists. However, a second look at the underlying data makes arguing for the relative or excessive fullness of the world quite ambiguous. First, the percentage drops to 25 per cent if aquatic NPP is taken into account. Second, it appears that adding up direct uses of biomass with land-use changes or 'diversion' (obtaining the co-option concept) and potential productivity losses from destruction reflects a confused concept of (non-)sustainability. As will be shown below in a consistent accounting system, land-use change may increase or decrease productivity and the environmental quality of natural resources. The direct use of NPP and the conversion of ecosystems to different uses might thus not necessarily detract from future human uses of the biosphere. Furthermore, and as pointed out by Vitousek et al. (1986), NPP appropriation does not represent an indicator of the global carrying capacity of the human population which would have to consider possibilities of technological advances in natural resource use and standards of living, i.e., consumption patterns of the projected population (see below, section 3.1).

The data on greenhouse gas emissions in Table 1.3 represent a 'business-as-usual' scenario, reflecting a doubling of CO_2 equivalents by the year 2030 and a global warming of about 3°C by 2100. While modellers seem to agree on this scenario, there is less consensus on what has happened over the past century: on one hand, it is argued that there is no unambiguous proof that the rise in temperature (of up to 0.6°C) is man-made (National Institute 1992: 142); on the other hand, it is maintained that the accumulation of carbon released to the atmosphere by human activity has exceeded absorption capacities of ecosystems, resulting in the 'seven hottest years on record in the last eleven years' (Goodland 1991).

Uncertainties also persist about the relevance of the business-as-usual scenario and regional patterns of climate change, and their effects and repercussions. Alternative scenarios have been presented by the Inter-governmental Panel on Climate Change (IPCC 1990) of:

1 business as-usual, assuming coal-intensive energy supply, modest increases in energy efficiency, modest CO_2 control and depletion of tropical forests;
2 large efficiency increase and stringent CO_2 control;

17

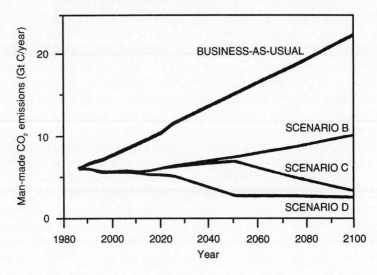

Scenario A (Business-as-usual) assumes that few or no steps are taken to limit greenhouse gas emissions. Energy use and clearing of tropical forests continue and fossil fuels, in particular coal, remain the world's primary energy source. The Montreal Protocol comes into effect but without strengthening and with less than 100 per cent compliance. Under this scenario, the equivalent of a doubling of pre-industrial CO_2 levels occurs by around 2025.

Scenario B (Low emissions scenario) assumes that the energy supply mix of fossil fuels shifts towards natural gas, large efficiency increases are achieved, deforestation is reversed and emissions of CFCs are reduced by 50 per cent from their 1986 levels. This results in an equivalent doubling of pre-industrial CO_2 by about 2040.

Scenario C (Control policies scenario) assumes that a shift towards renewable energies and safe nuclear energy takes place in the latter part of the next century, CFC gases are phased out and agricultural emissions (methane and nitrous oxide) are limited; an equivalent doubling of pre-industrial CO_2 will occur in about 2050.

Scenario D (Accelerated policies scenario) assumes that a rapid shift to renewable energies and safe nuclear energy takes place early in the next century, stringent emission controls in industrial countries and moderate growth of emissions in developing countries. This scenario, which assumes CO_2 emissions are reduced to 50 per cent of 1985 levels, stabilizes equivalent CO_2 concentrations at about twice the pre-industrial levels towards the end of the next century.

Figure 1.2 Projected man-made CO_2 emissions (billion tonnes carbon per year)
Source: IPCC (1990: 15).

3 a shift towards renewable energy sources and nuclear energy in the second half of the next century; and
4 a shift to renewables and nuclear power in the first half of the next century, stabilizing CO_2 emissions at 50 per cent of 1985 levels by the middle of the next century.

However, even the fourth – stabilization – scenario might involve reaching the limit of 2°C global warming indicated in Table 1.3 (National Institute 1992: 149). The different scenarios are illustrated in Figure 1.2.

Many uncertainties also surround the assessment of the depletion of the ozone layer and its effects on human health and ecosystems. Due to these uncertainties, sustainability standards cannot be expressed in terms of environmental or health impacts (*ibid.*: 131); at least for the time being, they have to be determined politically. If all the stipulations of the Montreal Protocol (see Table 1.3) were implemented, a concentration of 2 ppbv Cl at which no further depletion takes place would be reached by the middle of the next century (*ibid.*: 134). In the meantime, it is estimated that, in the absence of protective action and at a decrease of the ozone layer by 10 per cent for the middle latitudes, non-melanoma skin cancers will increase by 25 per cent and cataracts by 7 per cent. It is less clear to what extent ecosystems and especially bioproductive systems can adapt to changes in UV radiation (World Bank 1992: 63).

As indicated in Table 1.3, the qualitative degradation of agricultural land through erosion, salinization or waterlogging is probably a bigger threat to dryland areas than advancing deserts. Indeed, no clear trends of (advancing or retreating) vegetation cover could be detected from satellite imagery in the Sahel (*ibid.*: 55). About 11 per cent of the Earth's vegetated land has been moderately or severely degraded since 1945 (WRI 1992: 3). Figure 1.3 shows the distribution of this degradation over major regions in the world, notably Central America and Europe. The misery of environmental refugees in developing countries, suffering starvation, disease and disability, alerts to the ultimate effects of land loss and degradation in those countries. In order to pinpoint the underlying causes of these effects, further (up-dated) studies of the carrying capacity of lands need to be carried out (see Table 1.3). As is shown below (in section 5.2), such studies have the ability to assess the feasibility of sustainable development, especially in local 'ecoregions'.

Biodiversity refers to the range of ecosystems, species and genes in existence. It is probably the least known environmental concern both regarding its magnitude and its effects on human wealth and health. As shown in Table 1.3, estimates of species extinction, a clearly irreversible loss, vary greatly. Though not explicitly stated, annual extinction rates between 5000 to 150,000 (Goodland 1991) are anticipated if current trends of habitat destruction, overuse and pollution continue. Such rates are indeed 'mind-boggling' indicators of 'mass extinction' (Ryan 1992: 9). By contrast, past extinction rates are modest. The total number of species extinctions, since the seventeenth century, has been estimated at 724 (McNeely *et al.* 1990; cited in UNEP 1992: 54).

Table 1.3 indicates deforestation rates of close to 17 million ha annually in tropical areas. As these areas contain more than half of the world's biota, deforestation in these forests may cause the extinction of 15,000 to 50,000

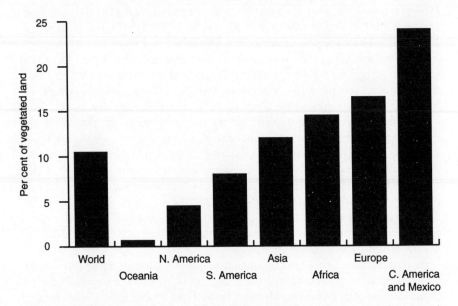

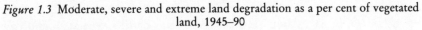

Figure 1.3 Moderate, severe and extreme land degradation as a per cent of vegetated land, 1945–90

Source: WRI (1992: 3).
Note: Figures for European and Asian parts of the Soviet Union are included with Europe and Asia.

species per annum between 1990 and 2020 (UNEP 1992: 54). Of course, this refers to only one of many other functions of forests, namely the provision of habitat for species. Other well-known functions include the maintenance of water regimes, soil preservation, absorption of CO_2, provision of fuelwood and timber and cultural values. The sustainability limits of wood consumption in Table 1.3 seem to indicate sustainable patterns of logging and consumption. Considering, however, the increasing non-accessibility of forests, differing regional patterns of exploitation and increases in world demand, it is anticipated that industrial demand alone for roundwood will not be met by the year 2010 (*ibid.*: 48).

Evidence given for running the world unsustainably in terms of energy consumption ranges between 50 years of availability of proven reserves of oil and gas and 800 years of estimated reserves of solids, oil and gas (see Table 1.3). Of course, other factors need to be considered, too, such as the use of renewable energy sources, possible increases in energy intensity in developing countries and possibilities of energy savings through substitution and improved efficiency of energy use. Increasing reserves of crude oil and gas and decreasing prices seem to indicate reduced scarcities in the availability of non-renewable natural resources (World Bank 1992: 37).

It is also debatable if the sink constraint of energy production and

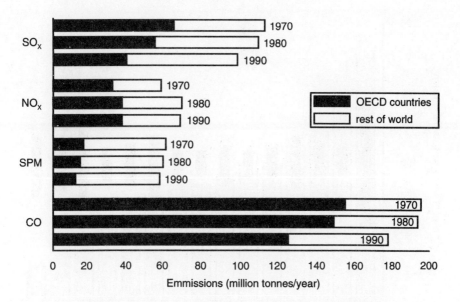

Figure 1.4 Man-made emissions of common air pollutants
Source: UNEP (1992: 1).

consumption is the 'key limit' (Goodland 1991) to economic growth, both in terms of emissions and effects on human welfare. Figure 1.4 indicates stagnant or decreasing trends in the emission of key air pollutants (with industrialized – OECD – countries still the main polluters of carbon monoxide). Accordingly, no clear trends in urban air quality in developing or developed countries can be detected. It appears, however, that, while some improvement has taken place in industrialized countries, ambient concentrations of sulphur dioxide and suspended particulates are frequently exceeding WHO guidelines in developing countries (UNEP 1992: 1–4).

Another global concern, not shown explicitly in the synoptic Table 1.3, is the degradation and depletion of oceans. While the open oceans still appear to be largely unaffected by human activity, pollution of coastal areas and enclosed or semi-enclosed seas has become a major international concern. In addition, world fish catch, rising from 60 million tonnes in 1979 to 91 million tonnes in 1989, seems to approach limits of 100 million tonnes set by the Food and Agriculture Organization (FAO) for avoiding the risk of substantial fish-stock depletion (*ibid.*: 29). In a regional breakdown by major fishing areas, sustainability limits seem to have already been reached in eight out of 17 fishing areas (Fig. 1.5).

Overconsumption in the North and underconsumption in the South may reflect a truly non-sustainable economic growth if this growth is to eliminate underconsumption of the poor by replacing it with the lifestyles of the rich.

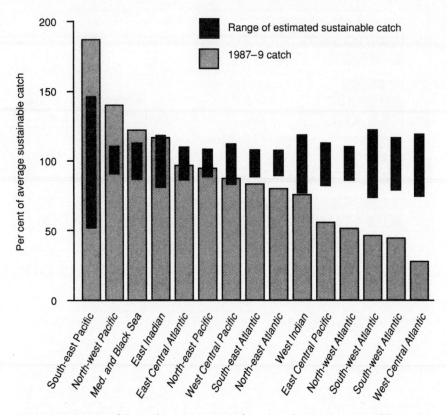

Figure 1.5 Marine fish catch as a per cent of estimated average sustainable catch, 1987–9

Source: WRI (1992: 17).

Note: Maximum sustainable yields are not defined for Arctic and Antarctic regions.

On the other hand, can we really expect the developing countries to content themselves 'forever' with a share of 15 per cent of world income (UNDP 1992b: 18)? Box 1.7 indicates that currently over 1 billion people live in poverty. Clearly, for those people and the countries in which they live, the elimination of poverty is a prerequisite for any action on global environmental concerns. As in the case of environmental trends, there is, however, a great deal of uncertainty about what production and consumption trends and patterns will prevail in the future. More sophisticated models (than the overly simplified limits-to-growth predictions) are needed to assess the feasibility of growth and development paths and patterns which would meet the immediate needs of the poor without depriving future generations of the use of scarce environmental assets.

The following examines further environmental conditions in poor

---

Box 1.7 Poverty in developing countries

Drawing the poverty line at less than US$370 per person per annum obtains 1.051 billion poor in 1985, projected to increase to 1.133 billion in 1990 and 1.107 billion in the year 2000. During this time period poverty is to decrease in Asia and to worsen in Africa, the Middle East and Latin America and the Caribbean. The projections are based on income growth and the assumption of a constant income distribution (World Bank 1992: 29, 30).

On the other hand, such constancy should not be taken for granted. UNDP estimates indicate that the considerable national income disparities (26 times between the richest fifth and the poorest fifth of people in the worst case of Brazil) are by far exceeded at the global level: there, the richest fifth obtain about 83 per cent of world income, while the poorest fifth have to make do with 1.4 per cent (UNDP 1992b: 3).

As shown in section 2.3, dollar figures cannot fully reflect human well-being or its absence. In fact, it appears that global disparities in terms of 'indicators of basic human survival' (primary education, life expectancy, infant and child mortality) have narrowed considerably (*ibid*.: 4).

---

countries. It is in those countries that the quality of life or life itself is directly linked to the quality of the immediate natural and man-made environment. The problems and opportunities for improving the human condition in harmony with environmental conditions are dramatically disclosed there.

Environmental problems in developing countries

The predominantly agricultural economies of developing countries depend on the availability and quality of natural resources for the sustained use of the bioproductive systems of agriculture, ranching, forestry and fishery. Degradation and depletion of land/soil, water and forests are the outstanding environmental problems of developing countries. The inability of local bioproductive systems to produce sufficient food for rapidly growing populations is the cause of perhaps the greatest scourge of humankind – hunger and malnutrition. The World Food Council (1991) estimated that by the end of the century about 650 million people, 95 per cent of whom are in developing countries, might be chronically hungry. Over 2 million deaths were caused by famine in the 1970s, according to an estimate by the Worldwatch Institute (Brown 1984: 188). Food production increased in the 1980s in Asia, but stagnated in Latin America and dropped in Africa, which also experienced the highest population growth.

Increasing world food demand exerts mounting pressures on land which is the major factor of production in agriculture. Figure 1.3 above showed that about one-quarter of vegetated land has been degraded moderately to severely in Central America and Mexico. Other developing regions experienced somewhat lower shares of 8 per cent in South America, 12 per cent in Asia and 14 per cent in Africa. As indicated in Table 1.3, 64 out of 117 developing countries might have insufficient land resources to meet their food needs by the year 2000, assuming that potential surplus food is moved only *within* countries. However, those countries would be able to support one and a half times of their projected (year 2000) population if no restriction to the movement of surplus food is assumed (FAO, UNFPA and IIASA 1982). Losses in grain output from land degradation have been estimated at 12 million tonnes annually, with soil erosion being responsible for two-thirds of this loss (Brown and Young 1990: 64).

The global stock of water is fixed; it can neither be increased nor diminished. Most of it (97 per cent) consists of ocean water. Less than 1 per cent is available for human use, since almost 80 per cent of freshwater is caught in icecaps and glaciers. Still, globally there is more than enough water to meet present and future demand. The problem is the distribution of the resource, which 'tends to be available in the wrong place, at the wrong time, or with the wrong quality' (United Nations 1978). Future water supply might become even more destabilized if current trends of deforestation continue (see Table 1.3).

Freshwater is an important factor of production in bioproductive systems. By far the largest part of current and projected total demand for freshwater is for irrigation. Water scarcity is expected to increase in developing countries, affecting both the consumption by humans and irrigation. World Bank estimates indicate that 22 countries, most of them in the Middle East, North Africa and Sub-Saharan Africa, have severely constrained renewable water resources of less than 1000 m^3 per capita and per annum (World Bank 1992: 48).

Irrigation frequently reduces soil productivity through salinization and waterlogging and causes the spread of water-related diseases such as malaria and schistosomiasis. Access to safe water and sanitation in developing countries has hardly improved during the last decade since increased supply of water and sanitation is largely offset by population growth. As a consequence, over 900 million people suffer from diarrhoeal diseases which cause the death of more than 3 million children each year (*ibid.*: 5).

Forests are a vital bioproductive system. They provide timber, fuel and habitat for a multitude of species (see Table 1.3); they also protect soils and regulate water balances and climate(s). In developing countries, tropical forests have declined by nearly one-fifth in this century as a result of agricultural land development and wood production for shelter and energy. Most of the wood in developing countries is used for cooking and heating.

Over 100 million people in developing countries are already experiencing acute firewood shortages. About 2.7 million will not be able to meet minimum needs on a sustainable basis by the year 2000 (Lanly 1982). For

24

these reasons, there is an urgent need to develop alternative sources of energy for home consumption. The consumption of commercial energy in developing countries is indeed increasing and is expected to 'dominate energy markets worldwide' (World Bank 1992: 114). Accordingly, much of the environmental impacts indicated in Table 1.3 will probably be 'imported' by developing countries in the future if no major breakthroughs in alternative energy source development can be made.

Natural events become disasters when they are the cause of losses in life, limb and property. The magnitude and frequency of disasters have increased dramatically between 1960 and 1990. Human and economic losses were concentrated in developing countries, compounding environmental impacts and constraints to economic growth and development. Over a million people were killed by natural disasters (including drought) during this period in developing countries as compared to some 30,000 in developed countries. This demonstrates the greater vulnerability of poor countries to environmental events (UNEP 1992: 59).

Conspicuous – and widely publicized – events such as volcanic eruptions, oil spills or nuclear accidents need to be assessed as to their long-term effects and underlying socio-economic conditions. For example, the largest eruption of the century – Mt Pinatubo in the Philippines – is expected to bring about a global cooling of 0.5°C over the next two to four years (*ibid.*: 61). The severity of impacts of earthquakes, tropical storms, floods and landslides has been aggravated by unsuitable habitation in slums, removal of vegetation buffers (for example, mangroves) and deforestation. The accidental spill of a toxic chemical at a pesticide plant in Bhopal (India) caused nearly 3000 deaths and 20,000 injuries. This is an indication that industrial accidents are no longer limited to developed countries but can be expected to increase in tandem with developing countries' efforts to industrialize.

Human settlements consist of physical elements, shelter and infrastructure, and of community services such as education, health, culture, welfare, recreation and nutrition (United Nations 1976: 37). In developing countries, the lack or the low quality of both the physical and the non-material components of human settlements are an illustration of how poverty affects the quality of the environment and of living conditions. Accelerated urbanization is at the root of most environmental impacts on human settlements in developing countries. This process seems to be the result of the natural growth of the urban population rather than of rural–urban migration (which is, however, a significant contributory factor) (Holdgate, Kassas and White 1982: 337, 338). It is estimated that by the year 2000 eight of the 10 largest cities will be in developing countries. Those trends seem to indicate a general 'failure of the Western development model': big cities are no longer a signal of successful industrialization, but rather an indicator of urban and rural decay and poverty (Otterbein 1991).

The most serious impact of urbanization in developing countries is the mushrooming of marginal settlements within and on the periphery of the large cities. In developing countries, about one-third of the urban population lives in overcrowded makeshift shelters which lack basic sanitation, clean water supply, waste-disposal facilities and adequate services of transportation, health, education and recreation (Holdgate, Kassas and White 1982: 342). Under these conditions, crime and prostitution flourish, and communicable diseases may reach epidemic proportions. Air pollution has now also become a serious problem in cities in developing countries. While air quality has generally improved in industrialized countries, it continues to deteriorate, especially with regard to SO_2 emissions, in developing countries (Adelman and Fetini 1990: 2).

Environmental diseases are those that are affected by living conditions, climate and water supply in developing countries. Average life expectancy is still considerably lower in those countries than in developed ones, but the difference is expected to narrow by the beginning of the twenty-first century. Communicable and vector-borne diseases such as respiratory diseases, malaria, schistosomiasis and onchocerciasis account for a high proportion of mortality and morbidity. Water-development schemes for hydropower, flood regulation and irrigation create and spread habitats for disease vectors. Infant mortality has been considered a more sensitive indicator of social and health conditions than life expectancy because small children are especially affected by mal- and undernutrition, poor sanitation and lack of medical care (Eckholm 1977: 22). Each year about 10.5 million children under age of 5 die from infectious and parasitic diseases in developing countries (UNEP 1992: 33).

1.3 Evaluation: how bad is it?

'Let the facts speak for themselves' has been the guiding principle for environmental data collection and dissemination of the World Resources Institute (WRI 1992: xi). But do they, and, if so, are they understandable? Even the above highly selective presentation of sustainability indicators presents a variety of data of different aspects of the same concern and in different units of measurement. Also, their relationship to qualitative predictions or quantitative limits of the sustainability of human activities is not always evident. As a consequence, some of the so-called facts might raise more questions than answers. For example:

- Are catastrophic consequences of climate change clearly indicated by so many centimetres of sea-level rises?
- Does the return to 1990 levels of greenhouse gas emissions increase or decrease the yields of bioproductive systems?
- What is the damage to our life-support systems from exceeding so many ppbV equivalents of chlorine in the upper atmosphere?

- Does 11 per cent of degraded vegetated land present a risk to food security?
- Do so many species lost annually compromise the ability of future generations to meet their needs? Of what?
- Are numbers of years in oil and gas reserves meaningful indicators of running the Earth unsustainably?

This list of questions, which refers to the global concerns described in Table 1.3, could be easily extended. Alternative indicators could be chosen, further global or local concerns could be introduced and differing criteria could be used to assess the sustainability of human activities at different regional levels. For years, politicians, researchers, the public media and the general public have been exposed to a multitude of tonnes, centimetres, ppbVs, cubic metres or hectares in what has been appropriately called information overload.

One way to deal with this overload would be to reduce the amount of data currently disseminated by selecting 'key' indicators, or combinations thereof in indices, to answer 'key' questions. Box 1.8 illustrates the arbitrariness necessarily involved in such selection, especially in the field of the environment. While data availability might very well be the principal reason for the imbalances in those 'balance sheets', there are near-insurmountable problems in obtaining agreement – both nationally and internationally – on what are the key questions and key indicators to answer them.

Any actual selection of indicators is thus bound to be subjective in nature. The subjective (normative) character of proposed sets of indicators should, therefore, be made more explicit. In principle, this could be achieved by setting clear environmental standards (limits, targets or thresholds) which would, in fact, permit answering the above question of how bad (or good) is it? As shown below, this is the approach taken to deal with the analysis of a broadly defined concept of sustainable development (see section 3.2). However, the development of environmental indicators and the establishment of widely accepted standards of sustainability is still more the exception than the rule at national as well as international levels.

As a consequence, the general neglect or 'denial' (see above, section 1.2) of these environmental indicators has been the result. Policy-makers have thus focused on those indicators that permitted the assessment of socio-economic progress in a nutshell – easy to understand and handle in political debates – that is, national product or income and their components of capital formation, consumption and value added. Those indicators succeed in assessing the results of a multitude of human activities, both beneficial (goods) and sacrificial (costs), through aggregation by means of a common *numéraire*, the monetary unit. Indeed, economic theory indicates that, at least under certain ideal conditions of perfect competition, the market prices of commodities and services reflect relative marginal utilities and thus individual preferences. In democratic societies, such influence of individual

---

Box 1.8 Balance sheet of human development (two examples from UNDP 1992b: 14/15)

Life expectancy (developing countries)

Progress	*Deprivation*
Increase by 17 years since 1960.	Death of 14 million children before age 5.

Life expectancy (industrial countries)

Progress	*Deprivation*
Average life expectancy is 75 years.	One in three adults smokes.
One doctor for every 460 people.	Nearly five people per thousand seriously injured.
Two-thirds of people eligible for public health insurance.	Cost of in-patient care increased by two-thirds since 1980.
	300,000 cases of AIDS.

Environment (industrial countries)

Progress	*Deprivation*
Energy intensity of production increased six times since 1965.	Greenhouse gas index is four times that of developing world.
Sixty per cent of people served by water facilities.	Forty-two kg of air pollutants emitted p.a. per 100 people.
Phase out of CFCs by year 2000.	Ten tonnes of hazardous waste p.a. per km^2.

---

decision-making is an appropriate criterion for evaluating the outcomes of economic production and consumption activities.

The main and widely applied instrument to measure and aggregate these outcomes are the national accounts. They have been recently revised, elaborating a more integrated system of the accounts of different sectors and institutions, including those of natural assets (Inter-Secretarial Working Group 1994). The next chapter will discuss possibilities for further modifying and extending those accounts in order to incorporate environmental costs and benefits into economic indicators. In this manner, it is hoped, environmental concerns can be presented to mainstream decision-makers together with economic ones in environmentally-adjusted aggregates that are as handy as the original (conventional) measures of economic performance.

The advantage of monetary aggregates is that they appeal to policy-makers because of their inherent evaluative capacity – making benefits and disbenefits (costs) of diverse economic activities directly comparable. There is, however, another, perhaps more important, substantive argument for using the integrative power of aggregation in monetary terms. This is the existence of interdependences among socio-economic and environmental conditions, processes and activities. Those interdependences were perhaps the most important finding of the report of the WCED (see Box 1.9).

Interdependences can also be identified from the above overview of the state of the environment and development. First, the preceding section showed that environmental pressures are caused by high-level economic growth as well as by the lack of it. Second, spatial interdependences can also be observed: global impacts can result from and affect socio-economic activities at the national level. For instance, the concern of climate change has its roots in the joint production of bads, i.e., unwanted residuals (emissions), together with the goods of economic production and consumption: most greenhouse gases stem from the use of fossil fuels by enterprises and private households. In turn, global warming, sea-level rise and changes in the quality of natural resources from pollution impair production and consumption activities.

As pointed out above, the fragmentation of responsibilities for economic policy and management of different natural resources, social and health policies, and environmental protection have effectively prevented an integrated approach that would take into account those interdependences. Departments of water, land, energy, or health, and peripheral environmental agencies dealt with their narrowly defined mandates, unable to cope with trade-offs or take advantage of synergisms through 'win–win' strategies. An integrated approach to all those manifold programmes and activities requires the evaluation of their benefits and costs, including those of their interaction with other sectors and actors in the economy and the environment. In principle, such evaluation can be achieved by valuation in monetary terms. A first step towards effective integration of economic and non-economic activities and processes is thus valuation, reflecting their relative importance in monetary units within a consistent framework. The capabilities

---

Box 1.9 Economic–ecological interdependence

We have in the past been concerned about the impacts of economic growth upon the environment. We are now forced to concern ourselves with the impacts of ecological stress – degradation of soils, water regimes, atmosphere, and forests – upon our economic prospects. We have in the more recent past been forced to face up to a sharp increase in economic interdependence among nations. We are now forced to accustom ourselves to an accelerating ecological interdependence among nations.

(WCED 1987: 5)

and limitations of integrated environmental and economic accounting in identifying and classifying relevant economic activities and environmental impacts and putting a price tag on those impacts are explored in the following chapter.

2

ACCOUNTING FOR SUSTAINABLE DEVELOPMENT

'Green accounting' has been in the limelight as a straightforward solution to integrating environmental and economic data and analytical variables. As will be shown in Chapter 3, those variables permit the operational definition of the concepts of sustainability and sustainable economic growth. The following examines to what extent interactive economic and environmental processes can be measured in environmentally-modified accounts currently being developed by the United Nations. The proposed System of integrated Environmental and Economic Accounting (SEEA) is outlined and critically reviewed, especially as to its capability of pricing environmental concerns and their effects on human well-being. First results of case-studies that test the proposed methodologies are also presented.

2.1 Environment and economic process

Conventional economics has treated the environment as external to its analysis, models and information systems. The above-described 'discovery' of interdependence of economy and environment calls for rethinking conventional economic thought. In fact, conventional economics has been blamed for perpetuating environmental deterioration through its neglect of the environment in mainstream policies and microeconomic decision-making of households and enterprises. Overconsumption of natural resources and overuse of environmental waste absorption capacities have been the result of this neglect.

Economics deals generally with transactions in scarce resources for purposes of production and consumption. Conventionally and conveniently, these transactions have been conceptualized and measured in terms of stocks and flows of those resources. Rethinking economics might thus usefully start by examining to what extent environmental concerns can be expressed in terms of stocks of environmental assets and flows of environmental services from the assets.

The economic exchange system comprises productive, consumptive and accumulative activities which create flows of goods and services (including

31

labour) between the system components. The environmental system consists of natural, man-made and social components. In Figure 2.1, the two systems are shown to overlap in the area of the man-made environment which is depicted as the result of capital formation. Emissions of wastes and pollutants, originating from production and consumption, are indicated by bold black arrows. Flows of natural resources to production and consumption are shown as broken arrows. These flows represent the two basic material (economic) functions of the natural environment: waste disposal from and resource supply to the economic system.

This presentation is based on the principle of mass conservation according to which resources may be used but are never consumed. All materials either return to the natural environment or are recycled into the economic process, generally in a modified form. Recycling is, therefore, the only ecologically sound way of waste disposal. Recycling is indicated explicitly in the flow chart by feedback loops of emissions. Destructive (disastrous) activities of the natural environment are symbolized as depletion flows from the man-made to the natural environment, and 'out of' the natural environment.

Environmental and economic activities may affect human welfare either positively or negatively. This is indicated in the figure by white arrows pointing from all economic and environmental components to the symbol for human beings. The consumption of useful goods and services provides individuals with what economists call utility or well-being. Toxic goods (or rather 'bads'), such as contaminating pollutants, impair human health. Active participation in the productive system may produce either positive or negative effects resulting from job satisfaction or stress. Marginal settlements are a source of much misery, while adequate shelter and other man-made facilities can contribute to a sense of security, well-being and self-esteem. The impacts of aggressive and altruistic actions of individuals and social groups are indicated as welfare effects of the social environment. And, though nature is a source of joy and recreation, it can also be a hazard in the case of natural disasters.

Many of those effects are not the objective of economic transactions but may nonetheless affect them, increasing or decreasing their costs and benefits. As indicated above, economic theory deals with those effects as externalities. Textbooks usually describe externalities as unintentional and thus uncompensated side-effects of production and consumption that affect the levels of consumer utility and enterprise costs. The logical approach to dealing with these external effects is to internalize them into the price–cost structure of the economy by means of taxes, subsidies or other market incentives. The resulting revised price system is expected to ensure the reallocation of resources in an optimal way. The polluter- and user-pays principles advocate the use of market incentives for environmental cost internalization. They reflect a philosophy of accountability of causing agents for their environmental impacts. Accounting for environmental costs and

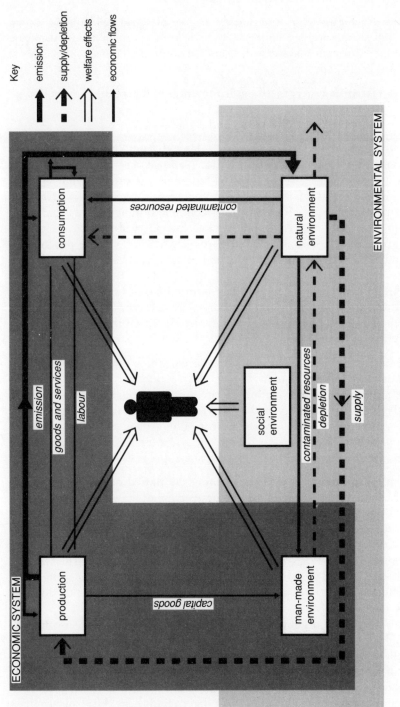

Figure 2.1 Environmental–economic system
Source: Bartelmus (1986: 8).

benefits is a means of quantifying the degree of accountability. As shown in the following, such accounting is at the heart of 'greening' the fundamental economic information system, the national accounts.

2.2 A system for integrated environmental and economic accounting

Objectives and framework

The structure of national accounts is derived from the elements of the economic flow system described above. National accounts measure those flows and are used in the analysis of economic performance and growth, and the assessment of the economic counterpart of social welfare. The above-described interactions between economic production, consumption, capital formation, the environment and human well-being call for the incorporation of environmental concerns and effects into the production, accumulation and income accounts and related balance sheets of tangible assets.

Conventional accounts have in fact been criticized for major drawbacks and distortions regarding the measurement of economic performance, growth and development (see, for example, Hueting 1989 or Repetto *et al.* 1989). These drawbacks include:

1 the neglect of scarcities of natural resources which threaten the sustained productivity of the economy;
2 the neglect of the degradation of environmental quality from pollution and other human activity, and its effects on human health and welfare; and
3 the accounting for environmental protection expenditures as increases in national income and product, even though those expenditures could be considered as social costs rather than social progress.

Some would even go as far as blaming the flawed national accounts at least partially for the degradation and destruction of the economy's environmental support systems: 'an expanding economy based on such an incomplete accounting system would be expected to slowly undermine itself, eventually collapsing as support systems are destroyed' (Brown 1993: 4). In the same vein, GDP has been considered as an 'evil indicator' (R. Hueting at the Special IARIW Conference on Environmental Accounting), misleading policy-makers into maximizing something that has little to do with utility and welfare.

However, those arguments seem to overlook that there is information beyond, or separate from, national accounts, for example, from environmental statistics and indicators (see above, section 1.3 and below, section 5.3), of which even hard-core economists are or should be aware. Just as environmental accounting is not a panacea for environmental problems, economic accounting is not the culprit that caused them. After all, accounts keep records only of what happened in the past. They may facilitate decisions but do not make them.

The recent revision of the United Nations *System of National Accounts*, the SNA (Inter-Secretarial Working Group 1994), presented a unique opportunity to examine how those flaws and drawbacks could be considered in modified accounts. In fact, the revised version of the SNA already includes in its asset accounts non-produced natural resources that provide economic benefits to their owners. However, so far, there is no international consensus on how to incorporate comprehensively environmental costs and benefits in national accounts. It appears, therefore, premature to radically change a well-established system of economic accounts that serves many different functions, in particular of short- and medium-term socio-economic analysis. On the other hand, the central framework of the SNA already covers various cost and capital items of natural resource stocks and use. This incorporation of environmental features in the internationally adopted SNA thus facilitates its use as a starting-point for environmental accounting.

UNCED, in its Agenda 21, requested that integrated environmental and economic accounting should be established in all member states of the United Nations and should be seen as a complement to, rather than a substitute for, traditional national accounting practices for the foreseeable future (United Nations 1993: para. 8.42). Detailed standards of environmental and natural resources accounting have therefore been elaborated in an SNA 'satellite system' rather than in the SNA core accounts. Satellite systems of national accounts serve to expand the analytic capacity of national accounting for selected areas of social concern without over-burdening or disrupting the core system (Schäfer and Stahmer 1990). The close linkage of a green satellite with its blue planet (the 'Blue Book' of the core SNA) should provide the appropriate remedy for the 'partial blindness' with which conventional economics and economists have looked at the environment (Gore 1993: ch. 10).

The Statistical Division of the United Nations (UNSTAT) has developed methodologies for a (satellite) System of integrated Environmental and Economic Accounting (SEEA) in an SNA handbook (United Nations 1993a). The system has been based on a framework for integrated accounting (Bartelmus, Stahmer and van Tongeren 1991) that was widely discussed by experts in the field, in international conferences and by the media (see, for example, *The Economist*, 18 January 1992). The present description of the main features and structure of the SEEA is based on this framework and the SNA handbook.

The following objectives of integrated accounting reflect responses to the above-mentioned drawbacks of conventional accounts:

Identification of all environment-related flows and stocks in conventional national accounts Environmental satellite accounts, in the narrow sense of detailed accounting for environmental expenditures and revenues, were pioneered by France (INSEE 1986b). There is now an increased interest in identifying all flows and stocks of assets in national accounts related to environmental issues and in estimating the total expenditure for the protection and enhancement of the different dimensions of the environment. One objective of this segregation is the identification of the increasing part of the Gross Domestic Product (GDP) which reflects the costs incurred to compensate for the negative impacts of economic growth on the environment, the so-called 'defensive expenditures' (Leipert 1989).

Incorporation of environmental costs and benefits into national accounts
The SEEA, in contrast to the more narrow approach to satellite accounting, provides a broader framework which covers also the 'external' and other imputed environmental costs and benefits. Taking the current state of knowledge and data availability into account, the system focuses on expanding and complementing the SNA with regard to two major issues, namely:

1 the use (depletion) of natural resources in production and final demand; and
2 the changes in environmental quality, resulting from pollution and other impacts of production, consumption and natural events, on one hand, and environmental protection and enhancement, on the other.

Possibilities of extending the SEEA for the analysis of environmental welfare effects, i.e., the 'damage costs' of the impairment of human health, recreation and other aesthetic or ethical values, are also discussed.

Accounting for the maintenance of national wealth   The general paradigm of sustainable development stresses the need to maintain the natural resource supply and waste-disposal functions of the natural environment. Environmental impacts that reflect the non-sustainable – wasteful or degrading – use of the environment are costed in the SEEA as indicated above. Those costs can also be viewed as a loss of functions of natural assets or 'natural capital' similar to the depreciation of man-made or produced 'fixed capital'. The SEEA extends, therefore, the concept of capital to cover not only man-made but also natural capital. It accounts thus for the maintenance of a broader concept of tangible wealth which includes natural assets. Correspondingly, the concept of capital formation is also expanded, introducing a concept of 'capital accumulation' which refers to the use of natural assets in economic production and consumption but may also include natural increase (growth) and discovery of natural resources.

Compilation of environmentally-adjusted macroeconomic indicators
The consideration of the costs of depletion of natural resources and changes

Table 2.1 Framework for integrated environmental and economic accounting (consolidated and modified)

			Final consumption		Capital accumulation			Rest of the world
					Produced assets			
Use/value added (Table 2.1.2)	Total	Domestic production (industries)	Households	Government	Except natural	Natural (biota)	Non-produced natural assets	Exports/ imports
Opening stocks					991.3	83.1	1744.4	
+ (plus)								
Use of goods and services	591.9	224.0	175.0	42.5	68.0	1.4	7.3	73.7
thereof: env. protection services	36.2	22.4	8.8	5.0				
Gross Domestic Product (GDP)		293.4						
Consumption of fixed capital		26.3			(23.0)	(3.3)		
Net Domestic Product (NDP)		267.1						
Use of natural assets (environmental costs)	(1.6)	58.9	17.1	(5.0)	5.1		(73.0)	(4.7)
Environmental adjustment of final demand (environmental costs)		17.2	(17.1)	5.0	(5.1)			
Environmentally-adjusted		249.8						
net Domestic Product (EDP)		191.0						
+ (plus)								
Supply (Table 2.1.1)								
Goods and services	591.9	517.4						74.5
Imports and residuals	(1.6)							(1.6)
+ (plus)								
Revaluation and adjustment to market valuation					138.1	13.5	464.0	
Other volume changes					(25.3)		22.8	
= (equals)								
Closing stocks					1149.1	94.7	2165.5	

Assets (Table 2.1.3):
Produced — Except natural, Natural (biota); Non-prod. natural asset

Source: Bartelmus (1992a).

in environmental quality permits the calculation of modified macroeconomic aggregates, notably an Environmentally-adjusted net Domestic Product (EDP). The possibility of introducing an Environmentally-adjusted National Income (ENI) concept by accounting for further welfare effects of environmental impacts and environmental protection is also discussed.

Table 2.1 illustrates the main features of the SEEA in a consolidated framework. The framework consists of three groups of accounting tables. The supply and use/value-added tables describe the flows of goods and services produced and imported (Table 2.1.1) and their use by economic production activities and final demand, i.e., intermediate and final consumption, capital accumulation and exports (Table 2.1.2). The assets accounts of economic and environmental assets (Table 2.1.3) show the stocks at, and changes of stocks between, the beginning and the end of the accounting period. Table 2.1 also illustrates that the stock accounts are linked to the flow accounts of the use/value-added table via accounts of capital accumulation. Those accounts thus form an integral part of both groups of tables. The main difference between this framework and conventional accounting lies in the introduction of environmental costs of depletion of natural resources and environmental degradation (largely from pollution). Those costs are mirrored in the expansion of capital asset boundaries to include corresponding changes in natural assets (shaded areas in Table 2.1).

The following provides a critical review of the proposed concepts and accounting procedures. This will facilitate an examination of how and to what extent those accounts can be used for measuring sustainable economic growth and development.

Accounting for defensive expenditures

Environmental protection has been considered as a typical example of 'defensive' rather than welfare or true income generating activities (see Box 2.1). Those activities are to defend society against a deterioration in environmental quality and its effects on human health and well-being. They aim at preventing or abating the undesirable side-effects of economic growth and urbanization and should thus not be considered as increases in social welfare. The additional deduction of defensive expenditures from National Income or product has therefore been suggested in order to calculate 'sustainable national income' or 'sustainable social net national product' (Daly 1989; Pearce, Markandya and Barbier 1990: 108). Box 2.1 describes a relatively broad concept of defensive expenditures as defined by its protagonist.

The subtraction of defensive expenditures from national accounts aggregates is controversial. The simple removal of those expenditures does not take into account the generation of value added in antecedent industries

---

Box 2.1   Classification of defensive expenditures (Leipert 1985)

- External costs of the general growth process of production and consumption (expenditures of environmental protection and compensation for damage caused by pollution).
- External costs of spatial concentration, centralization of production and associated urbanization (noise protection, rent increases, security and commuting costs).
- Increasing risk-susceptibility for the industrial system (provisions for hazardous industries, crime, defence, etc.).
- Other costs of car transport (accidents and emission control).
- Costs arising from unhealthy consumption and behavioural patterns as well as living and working conditions.

'Minimum' estimates of these items for Germany (excluding in particular the last item) amount to about 10 per cent of Gross National Product (GNP) 'during the last few years' (Leipert 1987).

---

which reflects the indirect cost of defensive expenditures. A more philosophical argument is that, in the final analysis, much, if not most, of the 'final bill of demand' could be considered defensive – against the deterioration of the current state of health and wealth of human beings (Bartelmus 1992a). Food consumption, for instance, could be considered as a cost of the sustenance of human life rather than an increase in well-being.

The SEEA confines itself, therefore, to identifying separately environmental protection activities by means of appropriate classifications. In Table 2.1, this is indicated by segregating environmental protection services from the total use of goods and services in intermediate consumption (of domestic production) and final consumption. In the SEEA, a further breakdown is made of domestic production activities producing environmental protection services as main, secondary or ancillary, i.e., non-marketed (internal), activities. A similar segregation is provided for the final consumption of households and government, and capital formation of produced assets. In the case of private households, an additional distinction between expenditures for direct environmental degradation and for further effects on human health and welfare is made.

The subtraction of defensive expenditure from output or value added as a social cost appears to be a confused approach to income and welfare measurement. On the other hand, the segregation of actual environmental expenses in national accounts provides valuable information on the level and composition of environmental action by industries, households, government and non-governmental organizations. This is the objective of more narrowly defined environmental satellite accounts, describing in greater detail 'incomings' and 'outlays' for environmental activities (INSEE 1986b).

Considering the proliferation of widely differing cost estimates of environmental measures, such accounting could facilitate the standardization of environmental cost concepts. By extension, such standards should also be applied in budgeting more accurately and consistently national and international action for sustainable development as proposed, for instance, in UNCED's Agenda 21 (United Nations 1993c).

In order to obtain a comprehensive picture of all environmental protection activities – whether carried out as internal activities of production units or as sales or deliveries to other units – one version of the SEEA proposes to 'externalize' internal (ancillary) environmental activities. Externalization is achieved by creating a new environmental service sector, thus increasing total gross output of the economy. National income and product are not affected, however, because the increase of value added in the new service sector is offset by a corresponding decrease in those sectors which originally contained the ancillary activities.

Accounting for sustainability in natural capital

As already indicated, cost allowances for the depreciation of produced capital are common practice in microeconomic (corporate) and national accounting. At the microeconomic level, such allowances are to permit reinvestment of funds set aside in order to maintain the capital base for sustained production. The introduction of another category of capital, i.e., natural capital, is to allow for further consumption of environmental assets in economic production and consumption processes. The concept of sustainability is thus broadened to include the maintenance of natural capital as well. As shown in Chapter 3, the maintenance of human capital of labour, skill and knowledge, and of institutional capital, providing the social, legal and organizational infrastructure for economic activities, should also be considered in a comprehensive analysis of the sustainability of production. Section 2.3, discussing the limitations of monetary valuation will show, however, that the monetary valuation of human capital and institutional support still meets with insurmountable problems.

The SEEA, therefore, concentrates on the incorporation of natural capital and changes therein. This is achieved by including natural assets in the asset boundaries. Table 2.1.3 illustrates this extension in shaded columns of stocks and (dis)accumulation of stocks of non-produced natural assets. In principle, *all* natural assets (see Box 2.2) are included in the system. In practice, however, in monetary accounts only those assets that can be assigned a monetary (economic or scarcity) value will appear. As a consequence, all those assets whose materials and services are freely available and are thus not 'scarce' (for example, in the sense of Walras 1926) would obtain a zero economic value and would thus not figure in the asset accounts.

Box 2.2   Natural assets, resources and resource accounting

The concept of *natural assets* has been developed and applied in the SEEA. It is broader than the concept of *natural resources*. Natural assets are close to what has been termed the 'natural heritage' or 'patrimony', including all environmental media (air, water, land) and ecosystems of value to society. Natural resources are usually defined more narrowly to refer to actually exploited assets, in particular for material inputs into economic production and consumption.

Natural *resource accounts* were pioneered in Norway (Alfsen, Bye and Lorentsen 1987). They present physical stocks of natural resources at the beginning of an accounting period, increases, decreases and adjustments during that period, and the stocks at the end of the period. *Natural patrimony accounting*, developed in France (Theys 1989), is broader, both in the sense of covering natural assets and in including, in principle, other – monetary – accounting approaches.

In line with the production boundary of the SNA, phenomena that take place wholly within the environment, i.e., outside the economic system, are therefore excluded. Such phenomena are probably better dealt with by complementary physical natural resource accounts (see Box 2.2) and systems of environment statistics and monitoring (see section 5.3, below). Also, effects from environmental quality degradation that affect human capital, i.e., human health and welfare, are probably beyond the reach of monetary valuation (see below, section 2.3).

Table 2.1 shows that the natural assets include both produced and non-produced assets. The former include naturally grown fixed assets, produced and used in (i.e., controlled by) natural resource-based economic sectors such as agriculture, forestry or fishery. They also include inventories (stocks) of naturally grown crops, livestock, timber or fish. Non-produced assets comprise the environmental assets, consisting of:

1 renewable resources 'in the public domain' such as marine resources or tropical forests whose growth is considered to be outside the managerial control of agriculture, forestry or fishing;
2 non-renewable resources of land, soil and sub-soil assets (mineral deposits); and
3 cyclical resources of air and water.

Underlying the monetary accounts of the SEEA are their physical counterparts, typically compiled in so-called natural resources or related patrimony accounts. Concepts and approaches of these – physical – accounts are highlighted in Box 2.2.

As indicated above, environmental assets are used as material inputs into production and final consumption, or as sinks for residuals (waste) from those activities. The impacts of depletion and (environmental quality) degradation on natural assets are costed in different valuation procedures described below (section 2.3). They represent costs of production and consumption. The costs of consumption are shifted, however, to the production sector as an 'environmental adjustment of final demand' in Table 2.1.2. Environmental costs are mirrored in the capital accounts as changes in the volume of available natural capital as part of a broader concept of capital accumulation (see Tables 2.1.2 and 2.1.3).

Changes in the availability of capital through capital formation (investment) or capital consumption are central to the analysis of economic growth and development. The modification of the accounting concept of capital formation into capital accumulation deserves, therefore, some elaboration. From the information contained in Table 2.1.3, the following concepts of capital accumulation can be derived, depending on the coverage of changes in the volume of natural assets:

1 capital accumulation in expanded production accounts;
2 capital accumulation, including natural resource discoveries and land-use changes; and
3 total capital accumulation, including the effects of non-economic events (disasters, accidents).

Box 2.3 provides details on the definition of those concepts with illustrative figures from Table 2.1.

'Nominal holding gains and losses' are not considered as volume changes in produced and non-produced capital. They do not affect capital accumulation, but are recorded as revaluation items of produced and non-produced assets in the asset accounts of the SEEA (see Table 2.1.3). Revaluation refers to changes in the market price of the different assets and is of particular importance in assessing the changes in national wealth in inflation-prone economies.

The above description of possible modifications of the concept of capital and – net – capital formation, due to the introduction of environmental assets, will facilitate the development and measurement of operational concepts of sustainable growth and development (see below, section 3.2). It might also prompt revisiting the key economic concepts of income and capital which appear still to mean different things to different scholars.

Environmentally-adjusted indicators: EDP and ENI

The accounting identities between income (value added generated) and expenditure (for final consumption and capital formation, modified by the difference between exports and imports) allow the calculation of Net Domestic Product (NDP) and Environmentally-adjusted net Domestic Product (EDP) as the difference between output and intermediate consumption, i.e., as the sum of domestic value added generated, and as the sum of

---

Box 2.3   Concepts of capital accumulation

Capital accumulation in expanded production accounts
The net amount of capital accumulation (CA) as shown in Table 2.1 consists of conventional 'produced' gross capital formation CF_p, reduced by capital consumption CC_p, plus an increase in non-produced assets CF_{np} (already recorded as capital formation in conventional accounts: mainly expenditures for land improvement and exploration of mineral resources) minus the 'consumption', i.e., depletion or degradation, of non-produced natural assets CC_{np}:

$$CA \ (net) = CF_p - CC_p + CF_{np} - CC_{np} = (68.0 + 1.4) - (23.0 + 3.3) + 7.3 - 73.0 = -22.6$$

CC_{np} is the key modification of national account, reflecting the environmental costs of using natural assets for purposes of natural resource inputs and waste absorption.

Capital accumulation, including discoveries and land-use changes
Another concept of capital accumulation is advanced in the SEEA which includes additionally 'other volume changes' *for economic purposes*, accounted for outside the production and income accounts. The reason for such accounting is to avoid major fluctuations in income and product aggregates due to unanticipated 'discoveries' (mainly of mineral resources) and changes in land-uses (for example, from forest land to agricultural to urban land). Discoveries amount to 27.8 in the detailed tabulations underlying Table 2.1 (Bartelmus, Stahmer and van Tongeren 1991: 131). They would turn a disinvestment of −22.6 into positive capital accumulation of 5.2. Land-use changes do not modify the overall aggregate of capital accumulation because an increase in one type of land-use (natural capital) category is of course balanced by a decrease in another one.

Total capital accumulation
Total capital accumulation includes further all other volume changes in tangible (non-financial) assets, due to reasons that are largely *non-economic* in nature. Such reasons include natural disasters, warfare, industrial accidents and (non-controlled) natural growth. Natural disasters and political events may affect both produced and non-produced assets. In the illustrative example, those effects amount to −30.3 and turn total capital accumulation again into a negative figure. As in the case of the 'economic' other volume changes, these 'non-economic' volume changes reflect capital gains and losses that could not be anticipated, disqualifying them from entering the production or income accounts (cf. United Nations 1977: 55). It has been argued, however, that exactly such 'windfalls' of capital gains (and losses) make national income 'a useful measure of real social income' (Hicks 1946: 180).

---

final demand categories. The definitory equations for these calculations, based on the figures of Table 2.1, are shown in Box 2.4.

EDP is thus derived either as the difference between NDP and environmental costs or as the sum of – environmentally-adjusted – final demand

---

Box 2.4 Definitions of NDP and EDP

Equations (1) show the value-added based definitions of NDP and EDP. The composition of NDP and EDP in terms of final demand categories is presented in equations (2).

(1a) $NDP = O - IC - CC = 517.4 - 224.0 - 26.3 = 267.1$

  where NDP = net domestic product
    O   = gross output
    IC   = intermediate consumption
    CC   = consumption of fixed capital

(1b) $EDP = NDP - EC = 267.1 - 76.1 = 191.0$

  where EC   = environmental costs

(2a) $NDP = C + CF - CC + (X - M) = 217.5 + 76.7 - 26.3 -$
    $0.8 = 267.1$

  where C   = final consumption
    CF   = capital formation
    CC   = capital consumption
    X   = exports
    M   = imports

(2b) $EDP = C + CA + (X - M) + (X_r - M_r) = 217.5 - 22.6 -$
    $0.8 - 3.1 = 191.0$

  where X_r = export of residuals
    M_r = import of residuals
    CA   = CA (net) as defined in Box 2.3

It should be noted that cross-boundary flows of residuals are recorded in Table 2.1 as negative imports M_r and exports X_r respectively, impairing the domestic environment by imports of residuals and relieving the environment by exports of residuals.

---

categories. The environmental costs (see Table 2.1) include the depletion and degradation of natural resources and environmental assets by domestic industries (58.9) and by final demand, i.e., by private households (17.1), governmental environmental protection (negative damage: −5.0) and scrapped capital goods (5.1). The definition of EDP in terms of final demand categories is obtained by adding the above-described capital accumulation component to the total of final consumption and the difference of environmentally modified (for cross-boundary flows of residuals) exports and imports.

The deduction of net transfers of income from abroad from NDP obtains National Income (NI). This indicator, especially per capita, is usually considered to be more welfare-oriented since it can be interpreted as the sum of monetary claims by economic agents on the available goods and services. It has therefore been suggested that the environmental welfare effects or 'costs borne' by households be deducted from NI to obtain an Environmentally-adjusted National Income (ENI) indicator (Boo *et al.* 1991; Bartelmus and van Tongeren 1993). No environmental adjustment of NI is proposed in the SEEA. SEEA focuses instead on productive uses of the environment and deals with welfare effects of these uses by applying a 'contingent valuation'. Such valuation assesses the willingness of households to forgo consumption for environmental protection (see below, section 2.3).

2.3 Pricing the priceless: the limits of monetary valuation

Methods of monetary valuation

Monetary valuation of natural capital is at the heart of integrated accounting. Without such valuation, aggregation and comparison of different economic and environmental activities and processes and their interaction is not possible. Conventional national accounts focus on market transactions, and valuation is thus based predominantly on market values. The advantage of this approach is of course the direct measurability of market values of transactions by means of statistical surveys. However, the use of natural capital in economic activities and further environmental cost and effects on human welfare cannot be measured in this manner. Those environmental uses and (dis)amenities are generally not bought and sold in the market, or at least not at full cost. For example, when using mineral resources, normally only their exploration or extraction costs are budgeted. The loss (consumption) of the natural capital stock of minerals is not accounted for, as is common practice in calculating a depreciation allowance for fixed capital consumption.

Introducing a monetary valuation for environmental goods and services facilitates choices between economic goods and services and environmental ones. This is what economics is all about (see, for example, Pearce, Markandya and Barbier 1990: 51 *et seq.*). Of course, the wisdom or rationality of those choices depends very much on the manner in which valuation is carried out in those areas where more or less anonymous supply and demand interaction is absent and prices are either controlled or have to be imputed. Any such imputations of values on the environment have therefore to be critically reviewed as to their underlying assumptions and methodologies.

A number of non-market or simulated market valuations have been advanced to assign a monetary value to both the stocks of natural assets and their use as material inputs and sinks for waste residuals. Three methods of valuation, *viz.*, market valuation, maintenance costing and contingent valuation, are distinguished and form the basis of three different 'versions' of the SEEA in *Integrated Environmental and Economic Accounting* (United Nations 1993a).

Market valuation of natural assets and asset changes   The use of market prices for measuring environmental depletion and degradation provides for consistency with the valuation of indicators compiled in conventional accounts. In the SNA, changes in market values of environmental assets due to depletion or degradation are accounted for as 'other volume changes' in conventional asset accounts. In the SEEA, by contrast, those volume changes are shifted to the income and production accounts, modifying NDP to obtain an EDP at market prices as shown above. As already indicated, this valuation applies only to 'economic assets' that have a scarcity value when used in production or consumption.

For those assets that are traded in the market, their current value can be determined by their market price and the ratio of their remaining lifetime over their total life-time, i.e., their replacement cost. In the case of fixed capital, the cost of using the assets in production is the depreciation allowance. Depreciation is based on the wear and tear (capital consumption) during the accounting period. Contrary to fixed capital assets, many natural assets are not traded due to public ownership or absence of property rights. However, where the products or services of those assets enter the market, prices are formed in trading those products. These prices can be used to determine the economic value of the asset, as well as value changes from its use. In principle, the economic value of a productive asset is derived from its intended purpose, i.e., the generation of income for its owner or user. In other words, the current economic value of an asset is determined by the expected (and discounted) value of the net profits resulting from the commercial exploitation of the asset.

Two valuation methods, based on the general depreciation principle of losses in income-spinning capacity of economic assets, have been advanced. The net-price valuation applies a unit value, consisting of the market price of a unit of the natural resource sold minus all factor costs actually incurred in exploiting the asset (Repetto *et al.* 1989). The user-cost valuation determines a cost allowance from the sale of the natural resource which, if invested, would generate a permanent (sustained) stream of income (El Serafy 1989). The different assumptions behind those methods are discussed in Box 2.5.

In general, the net-price valuation seems to be pessimistic, or at least neutral, about the potential reinvestment of the capital consumption allowance, deducting the total depletion value. This is similar to conventional depreciation accounting where the total replacement cost of capital is apportioned to the accounting period for deduction from GDP to obtain NDP. By contrast, the user-cost allowance is more optimistic, assuming that part of the revenue generated can be reinvested for achieving a permanent

46

Box 2.5 Valuation of natural resource depletion

The idea behind calculating a *user-cost allowance* for the depletion of exhaustible natural resources is to convert a limited stream of (net) revenues from the sales of the resource into a permanent income stream by investing a part of the revenue obtained from the sale of the resource, i.e., the user cost, over the lifetime of the resource (El Serafy 1989). Given a particular net revenue, the calculation requires only two parameters, the discount rate and the life expectancy of the resource. By contrast, determining the *net price* needs information on the market price and unit costs, including a normal return to capital. It can be shown that the user-cost method is a special case of defining depreciation as the change in the discounted value of a resource, for example, over one year of exploitation, assuming that the yearly net returns are constant for the remaining life of the resource (Hartwick and Hageman 1991). In turn, it is argued that under perfect market conditions economic depreciation is equal to physical depletion valued at the net price of the resource (Repetto *et al.* 1989). For a further discussion of those assumptions see Bartelmus, Lutz and Schweinfest (1992: annex IV).

(discounted) income stream equal to the total value of the resource (see also below, section 2.4). Another way to interpret those two valuations is considering the net-price method as valuing the use of an inventory of commodities that do not possess future income generation capacities, and the user-cost allowance as the use of services of a capital asset that is expected to yield income in the future.

Natural resource inputs enter the economic system as materials traded in the market. This permits the identification of a market price for those materials. However, such prices are usually not available for the services of the environment as a sink for pollutants and waste. The trading of pollution permits is not common practice yet. Only where the degradation of a natural asset affects directly its economic value, as for instance in the case of contaminated land, can changes in environmental quality of an asset be linked to its price (decrease). In that case, the value of the quality change can be expressed in principle in monetary terms. In practice, however, it will be difficult to distinguish such environmental quality changes from other quality and price changes.

Maintenance cost valuation Natural assets do not only serve as natural resource inputs into production. Environmental functions of waste absorption, genetic diversity, recreation and health, and the provision of cultural, aesthetic or spiritual values have been considered as 'services' of the environment that may be of even greater value than immediate economic (income-generating) uses of natural resources. As these services are not usually traded in markets, they do not fetch market prices, and market valuations cannot be applied. Their availability has, however, a 'social value' and their loss can be considered as a social cost. Their maintenance for present and future generations is desirable from a 'sustainability' point of view (see Chapter 3, below).

The costs of keeping those assets intact, either through restoration of depleted or degraded assets, or through preventive action that could have avoided depletion and degradation, are therefore estimated in the mainten- ance cost approach. Those costs are hypothetical in nature since actual depletion and degradation *have* taken place, and maintenance costs are estimated when in fact such maintenance was not carried out. Maintenance costs are thus potential costs of either achieving the quantity and quality levels of natural assets at the beginning of the accounting period or, if such levels cannot be 'reasonably' attained, at desirable environmental sustain- ability standards. Such standards would have to be set by social consensus or governmental fiat because of 'irreversibilities' of environmental effects or lack of maintenance techniques at acceptable (non-prohibitive) cost. The cost of meeting those standards could then be estimated for calculating the depreciation allowance at maintenance cost.

Contingent valuation The maintenance cost valuation focuses on the immediate origin of environmental stress from consumption and produc- tion. It measures the costs caused by those activities, during an accounting period, with a view to assessing the accountability of the different users of natural assets. An alternative cost concept would attempt to measure the costs borne ultimately by production units and private households. As to producing enterprises or establishments, diseconomies would have to be measured, in particular from pollution caused by other production or consumption units. As to households, externalities of production and con- sumption might impair their welfare, especially through health effects from contamination with pollutants. The cost-caused approach records the cost of environmental impacts during the period in which the initial impact was caused. By contrast, the assessment of the costs, ultimately borne by economic agents or natural systems, would have to record effects of environmental impacts from the past and current accounting periods that were ultimately borne by economic agents during the current period. For example, such effects could be the end product of a time-delayed sequence of emissions, ambient concentrations, exposure, contamination and final impairment of health.

The SEEA can be extended to account for those externalities borne by households and production units. Contingent valuations use statistical surveys soliciting the willingness-to-pay for the avoidance or mitigation of environmental effects. Other methods such as hedonic property prices, wage-risk studies or – increased – travel expenses have also been suggested especially in connection with cost-benefit analysis (see Box 2.6). The

---

Box 2.6   Cost/benefit valuation of environmental impacts

Various valuation methods for environmental damage or benefits have been proposed to assess the pros and cons of environmental and non-environmental programmes or projects (see, for example, Pearce, Markandya and Barbier 1990; and, for a more critical review, Gastaldo 1992). Methods of direct valuation, surrogate market values, potential expenditures and contingent valuation can be distinguished (Lutz and Munasinghe 1993).

Direct valuations assess

1   losses or gains in productivity from environmental impacts on the resource base of production (for example, losses of fish catch in polluted waters);
2   losses of earnings from health effects (including fatalities), for example, of contamination with pollutants;
3   actual defensive expenditures as a minimum estimate of environmental benefits (demonstrated willingness to pay).

Surrogate market values use price differentials between goods and services affected and non-affected by environmental impacts. They typically refer to differences in property values (hedonic price approach) but can also include travel costs (especially time differentials) and labour cost (wage differentials).

An example of *potential expenditures* are the above-described hypothetical maintenance costs, usually caused by different economic agents.

Contingent valuations attempt to solicit (for example, through surveys) the willingness to pay for potential environmental benefits or to accept compensation for losses of benefits (environmental damages), assuming (contingent on) the existence of a market for those benefits.

Integrated Environmental and Economic Accounting proposes a combination of market valuation with contingent valuation in one version of the SEEA which is based on the cost-borne concept. Contingent valuations would permit the measurement of the effects of environmental impacts on consumption as the amount private households would be prepared to sacrifice in order to avoid the deterioration of their environment. As already mentioned, the deduction of welfare-relevant costs borne by households from National Income for the compilation of Environmentally-adjusted National Income (ENI) has been suggested as an alternative to EDP calculation.

Inconsistencies and measurability

The three methods of valuation reflect to an increasing degree the distance of environmental processes and their effects from economic (production and consumption) activities. Problems of consistency of the valuations with – actually observed – market values and of data availability increase accordingly, as does their controversiality. The SEEA focuses, therefore, on the first two types of valuations or 'versions', based on – simulated – market values and maintenance costs, as widely applicable guidelines for environmental accounting. The third version, applying contingent valuation, and further versions, extending the production boundary of the SNA, are discussed only to provide a more comprehensive review of environmental accounting approaches, including those that qualify more for ad hoc modelling and research than for routine data compilation.

Market valuation in the SEEA is closest to conventional accounting. Changes in the market value of natural assets are recorded which are already accounted for in balance sheets of the SNA as 'other volume changes'. In the SEEA, those changes are shifted into the production accounts as environmental costs. As described above, these costs include the depletion of natural resources, as well as the degradation of those resources – to the extent that the environmental impacts are reflected in changed market values of those assets. The revised SNA makes allowances for such changes in its asset accounts, even if actual market prices are not observed and the stock or reserve values have to be estimated by their income-generating capacity, i.e., as (discounted) flows of net returns (rents).

Apart from simplifications for practical purposes (see Box 2.5, above), those valuations permit the calculation of environmentally-adjusted indicators which are consistent with the market values used in the SNA, from a macroeconomic point of view. From a microeconomic point of view, it can be argued, however, that these costs are imputed values that have not necessarily been internalized by individual economic agents (Bartelmus 1992a). Those costs have not gone through the mill of supply and demand interaction and corresponding price formation during the accounting period. Their deduction from value added generated or capital stock creates indicators that are not fully comparable with the 'observed' market values in conventional accounts.

On the other hand, full or partial internalization of environmental costs appears to be common practice in some of the high-risk or resource-dependent industries. Declarations to this end have been made in most environmental conferences involving industry representatives. For example, the Chairman of the Business Council for Sustainable Development predicted that 'environmental audits' will soon be the rule for companies seeking loans and insurance (Schmidheiny 1992). Recent examples of balance-sheet provisions for potential toxic waste clean-up are those of US chemical concerns (Monsanto, Du Pont, Cyanamid Cos), amounting to between US$200 and 400 million at the end of 1991 (*Wall Street Journal*, 23 March 1992). In those cases, corporate costing already includes environmental costs, and the prices formed in the market fully reflect those costs. Environmental (national) accounts would retrieve those costs from the residual ('catch-all') value of the operating surplus in conventional accounts, merely adjusting 'environmentally inflated' net value added to a more realistic value. Box 2.7 describes some principles of environmental accounting

---

Box 2.7   Corporate environmental accounting

Ideally, national environmental accounting should be supported by compatible corporate accounting as a basis for statistical surveys in this area. It appears that business spends more than ever on environmental protection, both for anticipating environmental impacts and for clean-up operations. Major companies are indeed prepared to be accountable for social costs to 'stakeholders' beyond accountability for conventional costs to 'shareholders' (Rubenstein 1989: 35).

However, accounting for actual environmental expenditures and potential environmental liabilities is not common practice yet and is largely left to the discretion of companies. In a recent survey of 222 transnational corporations only seven disclosed their environmental liabilities (United Nations 1992a: 1). Standard accounting methodologies need to be developed that disclose environmental expenses and liabilities and record the full costs of natural resource consumption, including its social costs. The work of an Inter-governmental Working Group of Experts on International Standards of Accounting and Reporting on (corporate) 'accounting for sustainable development', launched by the former United Nations Centre on Transnational Corporations, is a first step in this direction.

---

at the corporate level. Those principles have also been taken up by a United Nations project aiming at some standardization of corporate environmental accounting (see Box 2.7).

The market-value approach covers only those natural assets that have an economic value (in the SNA sense). It does not include environmental assets, such as air, wild land, waters and species, which are not included in SNA asset accounts. Nor can it account for all environmental functions of natural assets included in the SNA if those functions have not been reflected in the market value of the assets. The maintenance cost valuation attempts, therefore, to account comprehensively for both economic and non-economic functions of natural assets. In doing so, it remains as close as possible to the economic system, in order to ensure the measurability of environmental costs. A narrow cost-caused concept is applied: the costs refer to the immediate inputs (resource use) into and joint outputs (pollutants emitted) from production and consumption processes. The possible wider ultimate responsibilities for those environmental impacts of final users (domestic users and other nations) are thus not accounted for. They could (and should) be further explored in appropriate models of demand–production interaction.

Maintenance costing introduces further inconsistencies between market values and (maintenance) costs in the SEEA. As already indicated, those costs are not actually observed statistics but are estimates of the hypothetical costs of potential environmental restoration or prevention that did not actually take place. Such hypothetical costing or pricing is already applied in calculating the depreciation allowance for man-made capital where the replacements costs might in fact never be reinvested. The maintenance cost approach of keeping both man-made and natural capital intact seems thus to be consistent at least with regard to measuring capital services. Still, an inconsistency between the actual market price of a product, formed at the microeconomic level where environmental costs in general have not been internalized, and the use of the same price in macroeconomic accounts, incorporating environmental cost, remains.

One assumption to justify this approach would be that, if internalized, the environmental costs would not affect the product price. Of course, in reality, sellers would attempt to transfer most of the additional costs as price increases on to the buyers. The extent to which they are successful depends on prevailing price elasticities of supply and demand for the particular product. Accounting aggregates modified by hypothetical valuation are thus more to alert to structural distortions of the economy and unsustainable trends in its growth than to provide an accurate picture of past economic activity (Bartelmus 1992a). More importantly, the maintenance cost estimation permits the assessment of the – initial – level of market (dis)incentives required to prompt economic agents into internalizing environmental costs. Effluent charges, tradable pollution permits or user taxes have been proposed to bring about environmentally sound and sustainable production and consumption patterns (see below, Chapter 4).

On the other hand, the contingent valuation is applied to introduce a comprehensive cost-borne concept – not only by industry (through market valuation) but also by households, suffering the welfare (health, recreation, etc.) consequences of environmental impacts. Contingent valuations, based on opinion surveys of the willingness to pay for the avoidance of environmental hazards, have been applied with limited success in project/programme-oriented cost–benefit analyses. Their usefulness in *national* accounting is doubtful due to well-known problems (free-rider problem, short-sightedness of consumers, skewed impacts according to income distribution), which accumulate at the national level. Attempts at measuring national environmental damage in industrialized countries (summarized in Pearce, Markandya and Barbier 1990: 57–9) appear thus to use a mix of different valuations that do not clearly distinguish between cost-caused and -borne. As a consequence, the meaning of those aggregates becomes difficult to interpret.

Clearly, economic valuation of environmental effects, at the national level, reaches its limits when going beyond the cost/production-oriented approaches of market and maintenance valuation. Thus, earlier attempts at measuring economic welfare failed to replace or supplement conventional economic indicators (see below, section 2.4). Perhaps at the sub-national level, for instance for relatively small, well-researched 'eco-regions' – such as the Chesapeake project of the US Environmental Protection Agency (see below, section 2.4) – welfare-oriented accounting can be more successfully applied.

2.4 Case-studies

Correcting national output

A number of attempts at correcting macroeconomic indicators outside the national accounts have been made. Efforts to measure at least the economic counterpart of welfare by 'adjusted' or 'extended' national product figures (see, for example, Nordhaus and Tobin 1973) were made popular by Samuelson's textbook as Net Economic Welfare (NEW) (Samuelson and Nordhaus 1992: 429–31). Most of these approaches addressed two major shortcomings of conventional indicators of national product or income. One is the narrow economic boundary of those indicators, excluding household services such as food preparation, education and leisure activities, investment in consumer durables and disamenities of urbanization and environmental change. The other shortcoming is the treatment of outputs that do not yield any satisfaction in final consumption and should be considered, rather, as 'regrettable expenditures'. Those regrettables are close to what has been termed above 'defensive expenditures'. In the case of government, regrettables include expenditures for national security, which is 'defence' indeed, and the cost of maintaining a sanitary and safe natural and social environment. Household regrettables may include personal business expenses, commuting costs and expenditures for health and recreation effects of environmental degradation. The ambiguity of subtracting those expenditures from national income or product was discussed above (see section 2.2).

The magnitudes of estimates shown in Box 2.8 indicate the potentially great influence of any variations in concepts, scope and coverage of the adjustments and of changes in estimation procedures. As a consequence, even small variations in approach and data availability may affect the indicator more than actual changes in what it purports to measure. Together with conceptual, measurement and valuation problems, this may be the reason why NEW and similar welfare measures did not succeed in replacing national income and product as the standard economic aggregates.

Box 2.8   Net Economic Welfare (NEW), USA, 1965

For the USA, NEW per capita has grown more slowly than GNP. The slower growth is explained as probably 'inevitable' in a world that is becoming more congested and polluted (Samuelson and Nordhaus 1985: 119). In 1965, NEW was estimated to be about twice the size of GNP. Major additions are the estimates for leisure of the same magnitude as GNP itself and household production of about half the GNP. Deductions for regrettables amount to 15 per cent of GNP. Various non-conventional allowances for capital consumption and 'widening' are about a third of GNP which is three times the values of conventional depreciation.

To some degree, this argument also applies to an attempt at modifying NEW, by excluding its more arbitrary estimations, especially of changes in human capital and leisure, and by expanding the coverage of natural resource exploitation and environmental damage. Those modifications are the reasons for replacing 'new' by 'sustainable' in an Index of Sustainable Economic Welfare (ISEW) (Daly and Cobb 1989). Because of its narrower definition ISEW is lower than GNP – contrary to the NEW indicator which also accounts for the benefits of leisure. Over the 1950–86 period, the ratio of ISEW over GNP decreased in the USA from about 70 per cent to less than 50 per cent. According to the authors, this reflects a long-term trend of growing income inequality, exhaustion of natural resources and lack of investment in a sustainable economy (*ibid.*: 455).

In recent studies of two developing countries by the World Resources Institute, the costs of natural resource depletion were deducted from Gross and Net Domestic Product to obtain a 'Net' Domestic Product (which is actually a modified Gross Domestic Product) and an adjusted Net Domestic Product respectively. The results of these corrections are summarized in Box 2.9. The studies of the World Resources Institute demonstrate how physical changes in natural resources can be quantified in natural resource accounts. In focusing on the depletion of key natural resources of forests, soils, fisheries and minerals, their underlying database is much more valid than the quite heroic estimates, for example, of the use of leisure time and of other welfare-generating household activities for the NEW indicator. On the other hand, the omission of other (environmental) functions of natural assets, notably of waste absorption, tourism/recreation and genetic diversity, is a major drawback in estimating the sustainability of natural assets. Also, the ultimate objective of those studies is less to provide detailed physical resource accounts but to adjust GDP or NDP. Since those calculations are largely made outside the national accounting framework, they are not able to trace the impacts of overall deductions back to causing activities and institutions. Such linkage is perhaps the most important objective of integrated accounting for accountability. Moreover, the opportunities of

---

Box 2.9 Modification of GDP, Indonesia and Costa Rica

In Indonesia, the average annual growth rate of 'Net' Domestic Product (4.0 per cent) appeared to be significantly lower during 1971–84 than the rate of GDP (7.1 per cent) (Repetto *et al.* 1989). In Costa Rica, by contrast, there is hardly a difference in the growth rates of Net Domestic Product (NDP) (4.9 per cent) and adjusted NDP (4.7 per cent) over the 1970–89 period (Solórzano *et al.* 1991). In the average (over the respective time periods), a depletion rate of 14 per cent of GDP was calculated for Indonesia and of 5.7 per cent for Costa Rica.

using accounting identities and balances for validity and consistency checks are lost in this manner (Bartelmus and Tardos 1993).

Integrated accounting

The SEEA aims at overcoming the drawbacks of limited adjustments of only GDP or GNP by accounting comprehensively for all environmental impacts and linking them to those sectors and economic agents that have caused them. Case-studies carried out by UNSTAT and the World Bank in Mexico (van Tongeren *et al.* 1991), Papua New Guinea (PNG) (Bartelmus, Lutz and Schweinfest 1992) and Thailand (Bartelmus and Tardos, 1993) tested the methodologies proposed by the SEEA. These studies showed not only the importance of comprehensively accounting for both natural resource depletion and environmental degradation but revealed also the limitations and inconsistencies in existing databases. Clearly, the SEEA is still at an experimental stage. The feasibility of its concepts and accounting conventions needs further testing in countries at different stages of development before any international standards can be recommended.

The studies also reveal the importance of a sound physical database as a prerequisite for valuation of environmental stocks and flows. In Mexico, a major problem has been the large diversity of available data, compiled by different public and private institutions for different purposes. Fragmentation rather than lack of data has been the major obstacle to adapting environmental and economic statistics for incorporation into the accounting framework. By contrast, in PNG, a country at an early stage of economic development, statistical and accounting capacities were limited. The study thus provided for a feasibility test for integrated accounting with limited human and financial resources and within a relatively short time frame. The study in Thailand served illustrative purposes mostly. It represents therefore a starting-point only for a comprehensive national programme on integrated accounting.

Table 2.2 presents the summary results for Mexico and PNG. As already indicated, much of the value of these studies lies in the sectoral breakdown of natural assets and environmental costs. Environmental costs were calculated in two steps. First the costs of natural resource depletion were compiled and deducted from NDP and sectoral value added to obtain what is termed EDP 1 in the table. In the second step, the more controversial allowance for externalities, i.e., the costs of environmental quality degradation, were additionally deducted to obtain EDP 2.

EDP 1 reflects the cost of oil depletion, deforestation and land-use effects in Mexico, amounting to about 94 per cent of NDP (in 1985). In PNG, the only non-sustainable depletion of natural resources resulted from the exploitation of mineral resources; the deduction of those depletion costs obtained an EDP 1, ranging from 1 per cent to 8 per cent during the period

Table 2.2 Comparison of conventional and environmental accounting – selected indicators: Mexico and PNG

| | Mexico (1985) | | | PNG (1986–90)[a] | | |
| | Conventional accounts | Integrated ('green') accounts | | Conventional accounts | Integrated ('green') accounts | |
		EDP 1[b]	EDP 2[c]		EDP 1[d]	EDP 2[e]
NDP	42.1 billion P	39.7 billion P	36.4 billion P	2760 million K[f]	2580 million K[f]	2526 million K[f]
EDP/NDP	–	94%	87%	–	92–9%	90–7%
C/NDP	83%	88%	96%	89–100%	93–106%	95–109%
Δ CAP (net)	4.6 billion P	2.4 billion P	–0.7 billion P	463 million K[f]	282 million K[f]	228 million K[f]
Δ CAP/NDP	11%	6%	–2%	12–20%	5–17%	3–16%
NDP/CAP	37%	10%	–	59%	–	–

Notes:
[a] Lowest and highest percentage.
[b] Accounting for oil depletion, deforestation (including forest fires) and land-use (excl. fish and other species depletion).
[c] Accounting for air and water pollution, soil erosion, groundwater use and solid waste disposal; avoidance cost valuation.
[d] Net-rent valuation of mineral resources depletion.
[e] Potential damage restoration or avoidance cost valuation in the cases of waste water discharge (from mining); compensation cost for environmental impacts of forest clearing and dam construction.
[f] For 1990.

Sources: Mexico: van Tongeren et al. (1991); Papua New Guinea: Bartelmus, Lutz and Schweinfest (1992).

1986–90. Allowing for further costs of environmental quality degradation increased the percentages of EDP 2 over NDP to 13 per cent in Mexico (1985) and to 3–10 per cent in PNG (1986–90).

The net-price method rather than the user-cost allowance was applied for estimating the depletion allowances in Table 2.2. Compared to the net-price valuation which values actual resource depletion aiming at *capital* maintenance during the accounting period, the user-cost approach aims at *income* preservation (see above, section 2.3 as to the interpretation of the two valuation methods). Alternative calculations showed that the user costs are generally lower than the net-price based estimates (ranging between 12 and 46 per cent of the net-priced data in PNG). This is not surprising, considering that the user-cost method divides the net return from sales into a user-cost and a true-income component.

Environmental costs are mirrored in the SEEA in volume changes of capital (dis)accumulation. Table 2.2 shows that depletion and degradation of natural assets reduced their value, i.e., lowered net capital formation (ΔCAP, net) to nearly half their value in Mexico (60 per cent in PNG) in the case of natural resource depletion. If all environmental costs are taken into account an actual 'disinvestment' of −2 per cent can be observed in Mexico while capital accumulation is reduced to less than half in PNG. Corresponding increases of the share of final consumption from 83 per cent (C/NDP) to 96 per cent (C/EDP 2) in Mexico and to over 100 per cent in PNG illustrate dramatically a non-sustainable consumption pattern of living off the physical (natural and man-made) capital base. Especially in PNG, where physical environmental impacts have frequently been belittled as being local and small-scale in nature, integrated accounting demonstrates that the sum of those impacts may obtain a significant value – both relative to the causing activity and to overall national product.

Reductions in capital productivity are reflected in changes of the overall capital–output ratios (of NDP or EDP over CAP). Data on total capital stocks were available in Mexico only. They indicated an overall reduction in capital efficiency from 37 per cent to 10 per cent resulting from natural resource depletion only. Considerable fluctuations among the different economic sectors indicate a quite different picture of capital efficiencies if natural capital is used and accounted for in different sectors and production processes.

As pointed out above, welfare-oriented accounting approaches attempt to assess the environmental costs borne and environmental benefits enjoyed. In particular, those approaches make use of contingent valuations. This – highly experimental – valuation might be usefully applied to small regions. At that level, a limited number of inhabitants could be interviewed regarding their preferences for environmental services and dispreferences (dislikes) for environmental damages. Table 2.3 shows the results of such an experimental accounting study for the Chesapeake Bay, bordered by the US states of Virginia, Maryland and Pennsylvania.

Contrary to the above-described SEEA-based integrated accounting studies, the production boundary as defined in national accounts is explicitly extended in this study. Nature's 'production' of environmental and waste-disposal services is included in a separate account. As shown in Table 2.3, this is achieved by introducing a nature sector account which provides estimates of nature's outputs of final demand services and disposal of air and water pollutants, balanced by inputs of environmental damages. Contingent valuation seems to have been applied when valuing environmental damages. This would be the value inhabitants of the Bay area are willing to pay to avoid those damages. On the other hand, a potential (pollution control) cost valuation seems to have been used in assessing the value to polluters of the waste-disposal services (Peskin 1989; Grambsch, Michaels with Peskin 1993).

Table 2.3 Nature sector accounts for Chesapeaka – 1985 (millions of 1987 dollars)

Input		Output	
Environmental damages		Final demand for non-marketed	
Air	109.5	environmental services	
Water	346.8	Beach use	253.3
Total	456.3	Boating	140.1
		Recreational fishing	41.1
		Hiking	184.6
Net environmental benefit	839.7	Camping	159.5
(Disbenefit)		Waterfowl, deer and small	
		game hunting	129.9
		Wildlife observation,	
		photography and feeding	193.6
		Total	1102.1
		Waste-disposal services	
		Air	104.4
		Water	89.5
		Total	193.9
NATURE SECTOR INPUT	1296.0	NATURE SECTOR OUTPUT	1296.0
Environmental depreciation (−)	77.5	Environmental depreciation (−)	77.5
NET NATURE SECTOR INPUT	1218.5	NET NATURE SECTOR OUTPUT	1218.5

Source: Grambsch, Michaels with Peskin (1993).

It is interesting to note that to date, apart from some very tentative studies (for example, in Italy: Giannone and Carlucci 1991; or in Australia: Young 1991), most – monetary – environmental accounting has been carried out in developing countries. This may be partly due to United Nations efforts promoting integrated accounting in those countries; however, it might also reflect the information needs of natural resource-based economies of developing countries.

Industrialized countries are clearly more concerned about pollution problems that are more difficult to measure and value, as discussed above.

It remains to be seen if the opinion of the German Scientific Advisory Council on Environmental Accounting is representative for the attitude of developed countries. The Council recently recommended that, for now, efforts should be concentrated on the collection of physical data. Monetary valuation should be carried out only to the extent that the respective methodological questions can be resolved; the calculation of an 'eco-social product' is thus deemed 'not yet feasible' (Report of the 63rd Annual Session of the German Statistical Society, Braunschweig, 21–25 September 1992). The above-described methodologies of integrated accounting, advanced by UNSTAT, seem to provide ground for reconsidering this verdict. In fact, just recently, in his Earthday (5 June 1993) speech, the President of the USA directed the Bureau of Economic Analysis, which prepares the national accounts, to develop green GDP measures in order to highlight the economy's use of natural and environmental resources.

3

ENVIRONMENT AND DEVELOPMENT – A MATTER OF SUSTAINABILITY?

Setting out from the generic provision of economic goods and non-economic amenities, criteria for their sustained supply and use are explored. Sustainability of supply, use and users can be thus distinguished. The above-discussed integrated accounting system permits the description of the sustainability of supply in terms of capital maintenance. Correspondingly, sustainable economic growth is defined operationally in terms of environmentally-adjusted accounting indicators. Limitations of monetary valuation prompt the search for non-monetary yardsticks of multidimensional development. The sustainability of development is analysed in terms of feasibility of development programmes within a normative framework of targets, standards and thresholds. Finally, the relationships between sustainability and optimality criteria in economic analysis are discussed.

3.1 Sustainability criteria

Judging from the flurry of publications on the sustainability or non-sustainability of economic growth and development, it might appear that the fundamental optimality principle of conventional economics is being dethroned by the sustainability criterion of a new 'eco-nomics' (Postel 1990). The quest for sustainability seems to confirm the somewhat alarmist idea that 'ultimate limits' of economic growth are about to be reached. As a consequence, sustainability concerns would override allocative optimality of conventional economics (Daly 1991c). Such optimality refers to the best, most efficient and most beneficial, distribution of scarce resources among the producers of goods and services and, in turn, of those goods and services among final users or consumers (cf., for example, Samuelson's textbook definition of economics, in Samuelson and Nordhaus 1992: 3).

Chapter 1 discussed evidence, in terms of physical indicators, for the perceived non-sustainability of economic growth. The difficulties of evaluating those indicators ('how bad is it?'), due to their non-comparability among each other and with the benefits of economic activities, cast doubts, however, on the significance of the evidence given. Monetary valuation was therefore explored in an integrated environmental and economic accounting framework. The usefulness and limitations of such accounting, and supplementary or alternative non-monetary indicators in assessing sustainable growth and development is examined in the following.

Both affluence – through overconsumption – and poverty – through overuse of vulnerable natural resources for subsistence – are blamed for environmental deterioration in developed and developing countries respectively. As a result, natural resource stocks are depleted and vital life-support systems are impaired. Driving forces behind this overuse of natural assets are population growth and concentration and the application of environmentally damaging production and consumption practices and technologies. Obviously, there is quite a variety of sustainability notions, differing according to the basic questions of sustainability of what, for whom, where and when. Before dealing with the further question of 'how' in the next chapters, the meanings and interrelations of the different sustainability concepts and their relevance for economic growth and socio-economic development need to be clarified.

Figure 3.1 places the above-discussed indicators of (non)sustainability into a generic process of supply and use of economic and non-economic goods and services. Welfare (utility)-generating goods and services are shown to originate from the economic system, from nature made available as 'free' amenities or from the social system providing non-economic values and amenities to society. Variables or indicators that describe the flows of goods and services from their origin to final uses and users are indicated in brackets. Process or sustainability factors that may influence supply, distribution and use and may thus also affect the sustainability of the flow of goods and services are listed in shaded boxes. A distinction can be made between the sustainability of supply, uses and users of economic goods and services, and non-economic amenities according to the different stages of the supply–use process.

Sustainability of supply/production

Sustainability is originally an ecological concept, reflecting 'prudent behaviour' by a predator that avoids overexploiting its prey to ensure an 'optimum sustained yield' (Odum 1971: 223). In economics, the concept of income has been introduced as a similar 'guide for prudent conduct' in spending recurrent monetary receipts (Hicks 1946). Income is thus defined as 'the maximum value [one] can consume during a week and still expect to be as well off at the end of the week as . . . at the beginning'. The purpose is 'to give people an indication of the amount which they can consume without impoverishing themselves' (*ibid.*: 172).

Generalizing this notion for the whole population of a country yields a concept of national income as the amount a population can consume during

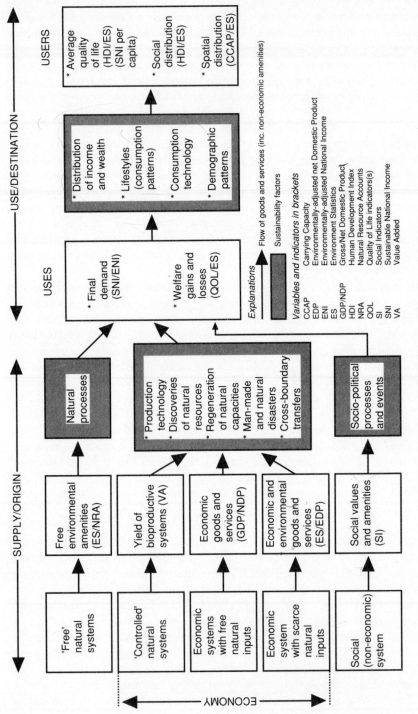

Figure 3.1 Sustainability and welfare generation

a period of time without impoverishing itself. In a closed economy, debts and credits among individuals cancel out in the aggregation of individual incomes. At a given level of a population and assuming otherwise rational behaviour, the main source of national impoverishment is to consume physical (produced) capital without replacing it or without reserving funds for alternative investment. A (cost) allowance for the depreciation of capital is, therefore, common practice in income (profit) accounting of production units. Depreciation is also recorded as a cost item in national accounts: it enters the gross value of production but is excluded from the contribution of the unit to national income, i.e., net value added.

Recent or recently discovered scarcities in further capital inputs of natural resource supply and environmental services of waste absorption have prompted the extension of the sustainability criterion from produced capital maintenance to the maintenance of natural capital. Focusing thus on basic input limitations or scarcities of the availability of production factors and their maintenance, different sustainability categories can be determined as a necessary condition for continuing economic production and growth. In addition to the maintenance of produced and natural capital, the maintenance of human capital of labour, skill and knowledge and of institutional capital, providing the social, legal and organizational infrastructure for economic activities and conflict resolution, could also be introduced in a comprehensive discussion of the sustainability of economic production.

The notion of sustainability of human capital or labour is controversial, since it is hardly possible to distinguish sustainability of economic activities of human beings from the general sustenance of human life. The quality of human life in its many material and non-material facets bears upon both the economic contributions of human beings and on their pursuit of non-economic goals. Furthermore, attaching an economic value to human beings is repugnant as it seems to imply their possible tradability. Consequently, the sustainability of human life is probably better analysed in reference to a set of standards of living or of the 'quality of life' as part of a broad multidimensional development concept.

A recent empirical–historical study of the role of institutions in development and environment indicates that institutions and institutional change are critical for the nature and speed of development, though not sufficient for stimulating economic growth (Adelman *et al.* 1992: 106). However, the flow of 'services' from the institutional set-up and its maintenance is even more difficult to conceptualize and value. This applies especially for 'disservices' from illegal and corruptive activities.

As discussed in Chapter 2, integrated accounting concentrates, therefore, on produced and natural capital consumption or use. Such accounting permits the replacement of conventional indicators of economic performance and growth, notably GDP or NDP, with the 'more sustainable' indicator of EDP. EDP thus reflects the scarcity of natural and produced capital inputs into the economic system and, consequently, a significant aspect of the sustainability of supplying goods and services, i.e., of economic sustainability (see Fig. 3.1). The step from economic production to the supply of non-economic, but welfare-relevant, amenities and thus from economic growth to development is made in Figure 3.1 by introducing two further sources or origins of human welfare, nature and the social system.

Natural systems provide free amenities such as water, oxygen, nutrient flows, capacities of waste assimilation and other less tangible services of space or recreation/enjoyment to production and consumption. As long as these goods and services are not scarce, they do not affect the sustainability of economic production and growth, and human welfare. They have no economic value and are therefore not accounted for in economic accounts and balance sheets. Other (social) value systems might give them an 'existence value' due to their ecological, aesthetic or ethical attributes. Changes in their availability or quality, i.e., their non-economic ecological sustainability, could be assessed in non-monetary or physical terms through systems of environmental monitoring and statistics, including (physical) natural resource accounts (see Box 2.2, above and section 5.3, below). If aggregate indices of the availability of these ecological assets can be found, ecological sustainability could be defined as non-decline of these indices. To the extent that changes in existence values reflect scarcities in their availability for economic uses, i.e., their economic sustainability, they can be given a monetary value in the sense of costing the maintenance of the natural systems. Such costing was discussed above (section 2.3) as one of the key valuation methods of integrated accounting.

The second source of non-economic welfare in Figure 3.1 is the social system. It includes public efforts to meet development objectives of equity, freedom, health, security, education, etc. Systems of social statistics and indicators attempt to measure the supply of economic and non-economic goods and services related to these objectives. Again, if aggregate indices of this supply can be determined, the sustainability of the provision of these goods, services and values, and, by proxy, of the social sustainability of development could be defined as non-decline of those indices.

As far as the supply of welfare-relevant goods and amenities is concerned, three categories of sustainability can thus be distinguished according to the source of supply, *viz.*, economic, ecological and social sustainability. However, various other factors may influence the supply/production process. In Figure 3.1, these sustainability factors are shown as shaded boxes, intersecting the flows of goods and services for purposes of graphical presentation. This presentation does not imply that these factors interfere only after the generation of final goods and services: rather, they can be inherent in the production/generation process (technology), appear before those processes are started (discovery of natural resources) or interfere with the final distribution process (disasters).

64

The sustainability factors can increase the scope for sustainability, for example, in the cases of alternative production processes and technologies or discovery of natural assets, or reduce sustainability through the disruptive effects of natural and man-made disasters. Possible alternative combinations of capital inputs that replace scarce (for example, natural) capital by less scarce (for example, human or produced) capital are the reason for distinguishing between strong and weak sustainability (Daly 1991c). Strong sustainability calls for the maintenance of each category of capital, assuming its complementarity (non-substitutability) in production and assigning, therefore, an 'existence value' for each capital category. Weak sustainability, on the other hand, focuses on maintenance of income or production, allowing for substitution and technological progress. It aims at the maintenance of the overall capital base, rather than of each of its components.

Sustainability of final use

For the analysis of the final destination of goods and amenities, a distinction is made in Figure 3.1 between

1 final uses in terms of final demand, needs and needs satisfaction, i.e., overall welfare generation; and
2 the ultimate receivers (users) of goods and amenities.

Such analysis is more welfare- and people-oriented than the above description of cost/technology-oriented production of goods and services.

As discussed above, National (disposable) Income (NI) represents potential claims on final uses of goods and services, and can be interpreted as an indicator of economic welfare. The incorporation into national income of further welfare effects from the degradation of the environment such as health effects or losses of aesthetic, cultural (natural/cultural patrimony) or even ethical (loss of our non-human 'life companions') values would facilitate a broader assessment of human welfare. The deduction of (monetized) losses in those values from national income obtains welfare-oriented indicators of Environmentally-adjusted National Income (ENI) or (more) Sustainable National Income (SNI). In other words, the sustainability of final demand could be defined in terms of non-decline of ENI or SNI.

However, economic goods and services and environmental effects from their production and consumption represent only part of the development story. Further welfare effects result from non-economic goods and services used for the satisfaction of other human needs and aspirations (see Box 1.1). In general, these goods and services are not or cannot be valued or costed in markets. Consequently, social evaluation will have to complement or replace monetary (market) valuation. Social evaluation can be carried out in principle by setting targets, thresholds or standards for non-economic development objectives.

Box 3.1   Selection and aggregation of social indicators

The usual procedure for determining social indicators is to specify social goals first and then to define sub-areas of objectives of a more instrumental nature. In turn, the sub-areas are to be represented by physical indicators (OECD 1973). Various procedures have been proposed to synthesize individual indicators into aggregate ones:

1   'Objective' or mathematical aggregation, ranging from simple summation or calculation of means to correlative reduction of variables to one- or two-dimensional indices (factor/component analysis).
2   Graphic superposition (overlay mapping) of indicators.
3   Direct measurement of 'compound' indicators with implicit weighting, for example, life expectancy which 'weights each disease in proportion to its effect in reducing length of life' (US Department 1961: 99).
4   Selection of exogenous weights on the basis of assumptions and hypotheses, or correlative analyses with other variables.

Simple mathematical aggregation procedures do not weigh indicators according to their relative importance, allocating implicitly equal weights to each measure. The same holds for the superposition of mapped indicators which can only give an initial insight into regional impact concentration. Factor-analytic calculations attribute weights according to the correlative relationships of the indicators. Compound indicators are limited in most cases to quite specific fields of analysis. Exogenous choices of weights usually reflect subjective values regarding the significance of the different indicators combined (Bartelmus 1980: 48, 49).

The policy focus on monetary measures of economic growth has, in fact, been criticized by advocates of multidimensional development. Such development would have to address a variety of social concerns or human needs as part of the overall goal of improving the quality of life. Much of the discussion around the notions of quality of life or levels of living and corresponding social indicators has therefore been on the inadequacy of a development process that concentrates on economic growth. Such development is prone to neglect social values of security, health, education, employment, distribution of income and wealth and environmental quality. In the absence of a common *numéraire*, social values are difficult, if not impossible, to aggregate. Of course, if an overall indicator of the quality of life could be found, the sustainability of human needs satisfaction could be defined as non-decline of this indicator. Box 3.1 describes the difficulties of aggregating indicators into more compound indices of the quality of life.

As in the case of supply/production, a number of sustainability factors, relating to consumption or use, may affect positively and negatively the long-term sustainability of final demand and human needs satisfaction. They include, in particular, changes in lifestyles, consumption techniques and the distribution of income and wealth.

Sustainability of users/populations

Indicators of sustainable final demand and human needs focus already to some extent on the people who strive to meet those needs. Indeed, the ultimate objective of sustainability is not to sustain human activity but human beings themselves. Although the sustainability of human and non-human capital in production is easier to measure, it should be considered as a proxy concept only for the sustainability of the welfare of human beings. This view reflects the philosophy behind human development which considers human beings as both means and ends of development (ul Haq 1989). However, this view could be seen as overly anthropocentric, neglecting the needs of non-human users of the biosphere. According to the above-discussed analysis of 'full-world economists', non-human users are indeed marginalized because of the human appropriation of an increasing share of the world's natural resources. This is also the view of 'deep ecologists' that deplore the isolation of human beings from Nature, considering themselves as 'superior to the rest of creation' (Devall and Sessions 1985: 65; see also below, section 3.3).

Attempts at concentrating more on the human factor in the assessment of development are indicated in the 'users' box in Figure 3.1. They include the

1 compilation of economic aggregates in per-capita terms;
2 definition of an overall human development index; and
3 assessment of the carrying capacity of land for human populations.

The trend of per-capita National Income (NI) or consumption is usually taken as an indicator of consumption/welfare-oriented economic growth – as compared to the more production/cost-oriented measure of economic growth, the Net Domestic Product (NDP). The sustainability of NI or ENI per capita would thus require a growth *rate* of the absolute value of these indicators that is at least equal to the population growth rate. However, like every average, per-capita income has little meaning unless assessed in the context of its distribution.

The consideration of distributionary factors raises a further development concern of equity in the distribution of the fruits of economic growth and the provision of non-economic amenities. Economic inequities are frequently associated with non-economic (environmental) ones. Inequalities within and among countries are aggravated by denying the poor and underprivileged access to increasingly scarce natural resources and by uneven spatial and social distribution of the impacts and costs of environmental degradation. For instance, gender bias in the distribution of work, income and wealth, especially in rural subsistence communities, has been blamed for depriving women of the results of economic growth and making them suffer the brunt of pressures on forests and croplands (Jacobson 1993).

The question is how to reflect all the different aspects of sustaining human beings in one single indicator. Social Accounting Matrices (SAMs) were designed as a modification of national accounts to measure income distribution over different institutions and social groups. The further incorporation of environmental effects into the SAM was also proposed (Boo *et al.* 1991). However, the above-discussed problems of measuring environmental welfare effects and allocating them to particular time periods and territories (see section 2.4) make those efforts rather academic.

The recently proposed Human Development Index (HDI) claims to be a measure of human progress by combining (averaging) National Income with two other indicators of adult literacy and life expectancy. Further adjustment for gender sensitivity and income distribution are also made, and the inclusion of a Human Freedom Index (see Box 3.2) is discussed (UNDP 1991). Assuming that the Index covers the most relevant development concerns and attaches correct weights to them, non-declining HDI would indeed reflect a sustained (human) development path. The question is if those assumptions really hold; if not, the composite Index would have to be broken up into and supplemented by sets of indicators whose combined performance could only be assessed in reference to explicitly stated targets or thresholds. As shown below, this would turn the sustainability analysis of development into one of the feasibility of development programmes complying with those targets and thresholds.

Finally, the sustainability analysis could focus directly on the (number of) people that can be sustained by a given territory, i.e., its carrying capacity. Such capacity depends on the desirable quality of life of the population in the territory, which would have to be defined in terms of essential or desirable standards of living. In this sense, carrying capacity analyses could provide an assessment of possible trade-offs between natural resource conservation and human needs satisfaction for purposes of ensuring a 'sustainable livelihood' (Desai 1991). In addition, the concept of carrying

---

Box 3.2 Human Freedom Index (HFI)

The HFI illustrates the problematique of constructing a seemingly objective yardstick of an intangible value-laden concern. The index is based on a – sometimes overlapping – listing of 40 rights to (for example, travel or consenting homosexuality), freedoms from (for example, torture or compulsory memberships), and freedoms for (for example, independent courts or equality for women). A simple presence/absence criterion is applied with equal weighting to each type of freedom/right. It would be even more arbitrary to weight the HFI within the overall HDI. It is probably for this reason that such inclusion has not been attempted. As could be expected, the HFI was hotly debated, especially by those countries at the lower end of the scale (Iraq, Libya and most centrally planned, socialist nations) (UNDP 1991: 20).

capacity of a territory for its inhabitants draws attention to non-sustainable trends in population growth, concentration and migration. However, apart from some ad hoc analyses of the capacity of selected bioproductive systems to feed people or to provide them with fuelwood (FAO, UNFPA and IIASA 1982; World Bank 1985), carrying capacity analyses have not been widely applied in development research and analysis. Perhaps, the focus of this ecological concept on sustaining non-human biological communities in their habitats could be usefully expanded to assess the potential of sustained local-level ecodevelopment of human communities (see below, section 5.2).

As compared to the sustainability of production, the sustainability of uses and users shifts the analysis from the maintenance of natural and man-made capital to the sustenance of people, regarding their monetary claims on goods and services, i.e., income, the satisfaction of their needs and aspirations and the supportive (carrying) capacity of their – local – environment(s). Such welfare-oriented approaches are difficult to assess in money terms. They have to resort to, or at least be complemented by, physical parameters, statistics and variables.

3.2 Definitions of sustainable growth and development

A – widely (re)cited – definition of sustainable development has been advanced by the WCED (1987: 43) as 'development that meets the needs of the present without compromising the ability of future generations to meet their own needs'. However, this definition is vague. It gives no indication of the time horizon ('future generations'), the scope and substance of human needs, nor of the role of environment (not even mentioned in the definition) in development. Differing interpretations and misunderstandings in using this definition might have been responsible for some of the suspicion of developing countries that the attribute 'sustainable' might introduce a new form of 'conditionality' for granting development aid.

Chapter 2 discussed in some detail more operational possibilities of merging environmental and economic costs and benefits in macroeconomic indicators. The preceding section also referred to various possibilities to go beyond economic integration (in money terms) by means of a non-monetary indicator or indicators of human development or the quality of life. Use will be made of those integrative measures in defining sustainable economic growth and development more rigorously and in operational (quantifiable) terms.

Sustainable economic growth

Integrated accounting permits the modification of the most widely used indicators of economic growth, GDP and NDP. It defines EDP which makes an allowance for the depletion and degradation of natural assets.

Consequently, the long-run increase or decrease of EDP can be taken as a measure of, or proxy for, a sustainable or non-sustainable path of economic growth. Focusing thus on the maintenance of produced and natural capital for ensuring continuing generation of output and value added and bearing in mind possibilities of extending the use of natural capital through (resource saving and environmentally sound) technological progress, resource discovery or substitution of produced, natural or human capital inputs, sustainable economic growth can be defined in operational terms as

upward trend of EDP, assuming that the allowances made for environmental depletion and degradation can and will be invested into capital maintenance and taking into account that past trends of depletion and degradation can be offset or mitigated by technological progress, discovery of natural resources and changes in production and consumption patterns.

The definition refers explicitly to the environmentally-adjusted indicator of EDP. The concept of sustainable economic growth differs, however, from the definition of this indicator as an essentially analytical (*ex-ante*) concept. The definition anticipates the above-described (see Fig. 3.1) sustainability factors of resource discovery, improvement in the efficiency of resource use, technological progress or changes in lifestyles. As a consequence, observed constraints of natural resource availability and environmental (services) capacities could be relaxed in the analysis of the sustainability of economic growth. Such anticipations lessen the operationality of the above definition, requiring predictive analysis, i.e., modelling. In addition, this definition assumes that environmental cost allowances are invested (rather than consumed) so as to ensure, for example, through substitution, importation or rehabilitation, that the total productive capacities of the economy are maintained.

On the other hand, sustainability of overall economic growth does not necessarily imply the sustainability of its components or determinants such as consumption, capital accumulation or the balance of foreign trade. Further effects of natural or man-made disasters, high inflation, indebtedness and changes in the institutional environment or in the productivity of human capital may of course also affect the sustainability of economic growth as defined above. Allowing for produced and natural capital consumption in the above definition obtains therefore only a 'more sustainable' growth concept that needs further refinement through more comprehensive models of sustainable economic growth.

The above concept of economic growth is production-oriented in its focus on domestic product and constraints in the quantity and quality of production factors; it reflects the 'costs caused' by production and consumption. However, economic growth analysis has also used national income, usually in per-capita terms, as its main operational variable, reflecting implicitly or explicitly a more welfare-oriented approach. Further negative welfare effects on human health, recreation and other cultural values from the deterioration of environmental quality are the ultimate concern of environmental protection. They can be interpreted as the 'costs borne' by human beings, detracting from the benefits of final uses of goods and services. As discussed above, integrated ac ounting could include those effects in further adjustment of private consumption and, consequently, EDP, or, alternatively, in indicators of ENI or SNI, per capita or in absolute terms. For purposes of welfare-oriented economic growth analysis, the above definition of economic growth could then be reformulated on the basis of those indicators.

A number of problems in assessing welfare-oriented growth make such definition rather ambiguous. First, like any average, per-capita figures are only meaningful if assessed in the context of their distribution. As pointed out above in the context of the sustainability of populations, distributionary aspects are particularly relevant for the assessment of equity in economic growth and development. Inequities are conspicuously manifest in conditions of poverty and hunger as compared to those of affluence, both within and among countries. However, they cannot be readily translated into economic costs or benefits. Second, the estimation of environmental damage, related to human well-being, faces considerable measurement and valuation problems which, in general, make a welfare-oriented definition of economic growth non-operational. Third, the reach of the monetized economy is frequently limited in developing countries to urban centres. Large areas in those countries are characterized by 'subsistence' activities that may exert significant pressures on natural resources and ecosystems at the local level. Alternative approaches to operationalize the sustainability of those aspects of human welfare are therefore explored below in non-monetary terms.

Sustainable development

Conventional economic analysis thus reaches its limits where the monetary valuation of non-economic effects of economic growth and of other human activities and natural processes become arbitrary

1 due to their remoteness from the economic supply and demand system;
2 where social concern overrides individual valuation implicit in market price formation; and
3 where the monetized economic system does not reach local-level communities.

A comprehensive concept of development has to cover a broader range of quantitative and qualitative aspects of the quality of life, i.e., human needs satisfaction. The notion of human development has therefore been taken up by UNDP (1992b: 2), defining it as 'the process of enlarging the range of people's choices - increasing their opportunities for education, health care, income and employment, and covering the full range of human choices from a sound physical environment to economic and political freedoms'. The definition responds to the need for expanding the analysis of development from focusing on its 'means (GNP growth)' to addressing its 'ultimate ends'. The focus on human well-being is also stressed by the Rio Declaration of the UNCED in its number one principle, stating that 'human beings are at the center of concerns for sustainable development' (United Nations 1993c). UNDP (1991) also proposes to measure human development by means of a Human Development Index. As discussed above (see section 3.1), it is questionable if the proposed Index succeeds in representing the main 'ends' of development and in reflecting correctly their relative importance.

It is therefore proposed here to operationalize those ends separately in terms of consumption levels or standards of living and non-economic targets and standards of an environmental, social, cultural or political nature. At the same time, such broadly defined development has to consider limits or constraints in the provision of need-satisfying goods, services and non-economic amenities that are not, or not fully, reflected in the cost calculations of microeconomic planning and decision-making. The transgression of those limits can be interpreted as 'social costs', reflecting society's evaluation of non-economic welfare losses from growth and development. The additional specification of limits in the availability of produced and natural capital permits the introduction of the physical counterpart of the sustainability of economic growth into development analysis. Minimum targets or standards and maximum limits or thresholds relevant in such analysis can be categorized as

1 standards of living or human needs satisfaction;
2 environmental limits of natural resource capacities and standards of pollution or contamination;
3 ecological limits of the carrying capacity of bioproductive systems or vulnerable lands for sustaining human populations in a particular territory;
4 standards of equity in the interpersonal, -regional and -generational distribution of income and wealth, access to natural resources, and the distribution of environmental costs and benefits; and
5 other cultural, political, social or demographic standards or targets.

The introduction of standards and targets in development analysis shifts the focus of sustainability from the maintenance of capital in economic growth to the examination of development programmes and activities regarding their compliance with minimum (desirable) standards of living and maximum standards of natural resource use and permissible degradation of the environment. Violation of standards or non-achievement of targets reflect a development pattern that *should* not be sustained in the long run. In this sense, the relatively neutral sustainability criterion of capital maintenance is replaced by a more normative one of the feasibility of development programmes.

The specification of targets, limits, standards or thresholds for sets of social goals and concerns is not sufficient for determining sustainable development. Trade-offs and synergisms among concerns and objectives occur because of interactions among human activities required for, and environmental processes interfering with, development. For example, a certain desirable level of an activity may not be compatible with ecological thresholds due to an excessive amount of jointly produced pollutants at that level. In this case, either the level of the activity or the thresholds would have to be adapted to the quandaries of reality in order to obtain a feasible level for this activity. As an alternative, an environmentally sound production process might become available and replace part or all of the original activity. On the other hand, synergisms in the improvement of social equity and environmental impacts could also occur. This would be the case, for instance, when alleviation of poverty reduces pressures of poor populations on marginal lands and congested cities.

The issue of environmental sustainability and soundness of development thus becomes a question of feasibility of development programmes or *the* development programme, operating within a normative framework of (exogenously set) targets, standards and thresholds. The determination of feasible programmes is a forward-looking approach which requires the modelling of future scenarios of activities and impacts. Assuming that such modelling can be carried out successfully and focusing on the basic objective of human (material and non-material) needs satisfaction and its environmental and social (distributionary) implications, an operational definition of sustainable (feasible) development can be advanced as

the set of development programmes that meets the targets of human needs satisfaction without violating long-term natural resource capacities and standards of environmental quality and social equity.

Interdependences among economic activities can be described by input–output matrices of the supply of goods and services. Similar interactions have been observed between the biotic and abiotic components of eco-systems, which again interact with human activities. If all those interactions can be formulated as relationships between single processes or activities which transform any kind of means (inputs) into ends (outputs), the powerful tools of input–output and activity analysis can be applied. Box 3.3 illustrates how feasible sets and levels of activities that satisfy constraints of minimum levels of human needs satisfaction ('inner limits') and maximum levels of production capacities, resource use and emissions of pollutants ('outer limits') can be determined (Bartelmus 1979).

The main drawback of the multidimensional concept of feasible development is its definition as a more or less large number of activities, the description of which requires an unwieldy amount of mostly physical data. Such information does not carry the same 'suggestive power' (Leipert 1984:

Box 3.3   Activity analysis of feasible development

The model referred to in the text can be expressed in matrix algebra as

$$A^*x \lesseqgtr \bar{c}, \bar{x}$$

where A^* is the inverse of the matrix obtained by subtracting the fixed linear input–output coefficient matrix A from the unit matrix I; x represents the activity levels, and $\bar{c}$ and $\bar{x}$ describe inner and outer limits respectively. The figure below illustrates a simple case of a two activity model of the production of food X_1 and shelter X_2, facing production capacity limits of $\bar{X}_1$ and $\bar{X}_2$, minimum requirements for food $\bar{C}_1$ and shelter $\bar{C}_2$, and maximum environmental limits for emissions of a pollutant $\bar{X}_p$ and of resource availability $\bar{X}_r$. The figure delimits a 'feasibility area' (shaded) of different combinations of levels of activities that satisfy human needs (of food and shelter) without violating capacity and environmental constraints. The introduction of pollution control by means of environmentally sound technologies would change the technology matrix A^* by decreasing in particular the pollution coefficients, thus shifting the environmental restriction $\bar{X}_p$ further outside for a possible increase in the feasibility area.

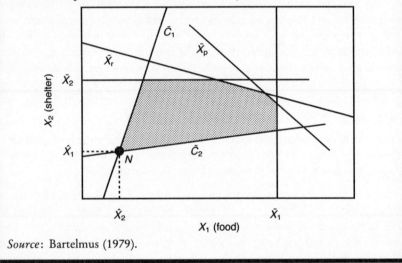

Source: Bartelmus (1979).

232) as aggregate monetary indicators. Also, physical data sets of 'levels' of activities do not reflect the relative importance of these activities as do monetary indicators: such indicators are weighted by the price system that, at least in approximation, reflects the interplay of individual preferences according to existing supply and demand patterns. The purpose of such weighting is of course aggregation for comparing the significance of a multitude of economic and environmental activities and processes according to their beneficial and detrimental effects. It has therefore been suggested that alternative development strategies be valued in terms of the costs

required to meet the standards and targets of these alternative scenarios. The total costs of meeting standards, for example, of natural resource use and pollution, could also be compared with National Income figures. The difference between these two aggregates would then 'show politicians and the public in monetary terms how far the country has drifted away from the course of sustainable economic development' (Hueting 1989: 38). A similar (but focusing on the direct environmental impacts during *one* accounting period) approach is taken in the proposed integrated accounting system by estimating imputed environmental 'avoidance' or 'restoration' costs (Bartelmus, Stahmer and van Tongeren 1991: 125).

3.3 Sustainability and optimality

Does the 'tide of environmental destruction' require 'a major overhaul of the rules, measures, and goals of our economic systems' (Postel 1990)? Is 'quantitative' economic growth obsolete – to be replaced by 'qualitative improvement' (Daly 1991c)? Should conventional economists be dying out (Daly 1991b)? In summary, are the optimization criteria of conventional economics overthrown by the sustainability criteria of a new 'eco-nomics' (Postel 1990)?

As shown above, operational sustainability criteria can be formulated for economic growth by extending conventionally measured capital consumption and maintenance to include natural capital. In this manner, an additional cost item of natural capital depreciation can be defined both for micro- and macroeconomic analysis. In microeconomics, traditionally unaccounted for costs and (dis)utilities have been treated as externalities. Theoretically, those external effects can be internalized into consumption and production functions by appropriate taxation. As a result (assuming conveniently shaped production and consumption functions), optimal, utility- and profit-maximizing, consumption and production patterns are realized. This concept of cost internalization has gained increased attention from advocates of market (dis)incentives designed to bring about efficient natural resource use and environmentally sound production and consumption. Its merits and practicability are discussed in Chapter 4 (section 4.3).

Much of the confusion and ambiguities in conventional economics sets in when microeconomic functions are – explicitly or implicitly – aggregated in macroeconomic analysis. Typically, the usefulness of macroeconomic aggregates and variables is judged in microeconomic terms, i.e., according to the extent they can reflect additions to or deductions from aggregate human well-being or utility. This is despite the well-known fact that such utility cannot be aggregated, nor can aggregate (sectoral and national) activities be unequivocally traced back to (and predicted from) individual behaviours. In other words, the relationships between microeconomics and macroeconomics are murky, but are used nevertheless to assess the – theoretical – merits of macroeconomic models.

From this point of view, much of the discussion of optimality vs. sustainability in economic growth and development seems to be quite irrelevant for actual policy-making. Definitions of sustainable development as non-declining human welfare (Pezzey 1989) or utility (Mäler 1991: 11) and models of achieving sustainable or optimal (maximum discounted) welfare (see, for example, Pezzey 1989; or Dasgupta and Mäler 1990: appendix) are thus probably more of academic interest. Of course, there is merit in (re)examining the theoretical foundations of conventional indicators and variables such as National Income or product, especially when emerging (environmental) concerns call for their modification. However, the assumption of optimality can indeed be seen as an 'embarrassing load' whose function is to justify using GDP as a welfare measure, advocating 'the semi-fiction that market prices accurately reflect scarcities' (Solow 1992: 10).

As already pointed out, non-sustainability of supplying goods and services, due to traditionally neglected capital scarcities, can be accounted for as additional costs of capital consumption in conventional economic analysis. Economic models could attempt to describe the most efficient (optimal) allocation and use of resources in terms of individual preferences (demand) and production capabilities (supply), reflected in – full-cost (including diseconomies) – shadow prices. Such analysis provides the justification for policies that aim at changing the (market) behaviour of producers and consumers by penalizing (taxing) wasteful and polluting activities and rewarding (subsidizing) environmentally-friendly activities.

The situation is different with regard to the sustainability of users as defined above in section 3.1. There, criteria of equity in the interpersonal, -regional and -generational distribution of economic benefits and environmental costs emerge. Allocative efficiency, i.e., a situation in which maximum output is produced with given resources, does not provide – even under 'perfect' market conditions – for distributional equity and thus social optimality (see, for example, Howarth and Norgaard 1990 in the context of intergenerational equity). In the absence of a social welfare function which would provide criteria for selecting a socially optimal set and level of human activities and resource (ownership) allocation, the only practical alternative seems to be the specification of normative targets or standards of the equitable distribution of produced and natural capital and the results of capital used (see above, section 3.2).

The question is to what extent efficiency criteria still hold when such standards are set and when feasibility criteria might overrule microeconomic (efficiency) optima. One author maintains that efficiency of resource allocation becomes irrelevant when economic growth becomes unsustainable: an overloaded boat will sink even if its load is optimally distributed (Daly 1991a: 35). What is called for is qualitative development rather than quantitative growth. Sustainability thus overrides efficiency.

This conclusion comes close to the tenets of deep ecology which considers

---

Box 3.4 Frontier economics vs. deep ecology

Dominant world-view

- Dominance over nature.
- Natural environment as resource for humans.
- Material economic growth for growing human population.
- Belief in ample resource reserves.
- High technological progress and solutions.
- Consumerism.
- National/centralized community.

Deep ecology

- Harmony with nature symbiosis.
- All nature has instrinsic worth; biospecies equality.
- Elegantly simple material needs (material goals serving the larger goal of self-realization).
- Earth 'supplies' limited.
- Appropriate technology: non-dominating science.
- Doing with enough/recycling.
- Minority tradition/bioregion.

Source: Devall and Sessions (1985: 69).

---

the preservation of ecosystems as an imperative that overrules any anthropocentric views of human needs satisfaction. A comparison of the 'dominant world-view' with the 'deep ecology' view is given in Box 3.4. The former reflects an attitude of 'frontier economics' (Boulding 1966) that assumes unlimited supply of natural resources and waste absorption capacities of nature. It also considers human beings as 'fundamentally separate from Nature'. By contrast, deep ecology 'attempts to articulate a comprehensive religious and philosophical world view' (Devall and Sessions 1985: 65). It stresses the intrinsic equality of species and condemns the exploitation of nature for human needs at the expense of non-human biota (*ibid.*; Naess 1976).

Such normative considerations can of course not be contested by economic (efficiency or cost–benefit) arguments. They should, however, not be easily dismissed either. Ethical or philosophical values do play an important role in the environmental discussion and, in fact, in political socio-economic decision-making. This is reflected in the above definition of sustainable development which takes non-economic objectives (standards) into account. Where the market is unable to cost properly the non-sustainability effects of the use or abuse of environmental assets, some kinds of standards for what presumably are the qualitative aspects of development need indeed to be set. As shown above, those standards can be seen as constraints for those activities that contribute to the non-sustainability of development. Thus, sustainability does not really 'overrule' optimality but – depending on possibilities of costing non-sustainability – can be seen as either an additional cost factor in economic performance and growth or as constraints that determine a feasibility space for development programmes.

Within such a space, i.e., within the normative framework of social standards, efficiency criteria still apply. It is only where microeconomic optimizing behaviour clashes with those standards that 'second-best' solutions (from the microeconomic point of view) would have to be found to comply with overriding social values. If such compliance does not take place, i.e., social standards are violated, the effects of such violation reflect the social cost of economic performance or growth. If those effects could be expressed in money terms, optimal performance or growth would be achieved where the marginal social benefits of economic growth equal the marginal social costs. Such calculations are hardly possible in reality. Perhaps a more pragmatic approach would be to apply exogenously set optimizing criteria to the feasibility space of the above-described activity analysis model of feasible development (see section 3.2 and Box 3.3). This can be done by means of an objective function which reflects existing prices or other weightings of activity outcomes. The least-cost or maximum-output value of the optimal set of activities can then be determined as the solution of a linear (or non-linear) programming problem (Bartelmus 1979: 266). The conditions for obtaining such a solution have been discussed, in the context of economic analysis, by Dorfman, Samuelson and Solow (1958).

In conclusion, efficiency, and thus common – optimizing – economics, provide still the primary criteria of using and producing scarce goods and services, modified only where social (governmental) fiat replaces individual preferences and decision-making. Considering actual experience with central planning and fiat, even regarding many of the qualitative aspects of development, such (over)ruling should probably be applied only sparingly (see Chapter 4, below).

4

PLANNING AND POLICIES I: SUSTAINABLE GROWTH AND STRUCTURAL CHANGE

The discussion of concepts and definitions of sustainable growth and development was to clarify the interrelations between environmental concerns and development objectives with a view to moving the environment into mainstream socio-economic policy-making. The following chapters examine policies and strategies that suggest themselves as a result of the conceptual discussion and that can be monitored by the operational variables defined above. Policies of sustainable economic growth and performance are discussed in the present chapter, and those of sustainable development in Chapter 5.

The present chapter introduces a strategic framework for the formulation and implementation of consistent policies at national, local and international levels. It explores the role of environmentally-adjusted macroeconomic variables ('eco-variables') in environmentally sound and sustainable policies of economic growth, employment, trade and foreign debt. The chapter concludes with reviewing the application of market instruments for environmental cost internalization and structural change.

4.1 Policy framework

Sustainable development has been the leitmotiv of the Earth Summit (UNCED) in Rio de Janeiro. It can be expected to remain on the international agenda during the twenty-first century. This at least is the purpose of the action programme of UNCED, its Agenda 21. However, in order to translate conference rhetoric into action some coherence needs to be given to what appears now just a long list of desirables in Agenda 21 and other national and international proclamations on sustainable development. Box 4.1 illustrates such a listing of 'policy directions', proposed by the protagonist of sustainable development, the World Commission on Environment and Development (WCED 1987). A flurry of more or less detailed recommendations, action programmes or proclamations on sustainable development by national and international organizations has been the result of setting in motion the preparatory process for the Rio Summit (see also

---

Box 4.1   Policy directions for sustainable development

1 *Population and human resources*:
  – elimination of mass poverty;
  – limitation of extreme population growth;
  – human resource development;
  – support of tribal and indigenous people.

2 *Food security – sustaining the potential:*
  – increase of food crops in developing countries;
  – reduction of food surplus in industrialized countries;
  – promotion of ecologically-sound farming;
  – land reform and protection of vulnerable subsistence activities (integrated rural development).

3 *Species and ecosystems – resources for development*:
  – increases in protected areas;
  – establishment of a 'Species Convention'.

4 *Energy – choices for environment and development*:
  – energy efficiency policies;
  – research and development: environmentally sound and safe energy alternatives;
  – 'conservation pricing' of energy commodities.

5 *Industry – producing more with less*:
  – export control of hazardous chemicals;
  – control of hazardous waste dumping;
  – technology assistance to developing countries.

6 *The urban challenge*:
  – settlement strategies for urbanization control;
  – decentralization policies.

7 *International co-operation*:
  – technical assistance for environmentally sound projects and policies;
  – incorporation of environmental concerns into international commodity arrangements;
  – disarmament and use of peace dividend.

Source: WCED (1987).

Chapter 6). They differ in focus, detail, scope and institutional responsibility, but seem to cover similar ground as the WCED. As a consequence, there is a 'risk of suboptimization' (Archibugi 1992: 16–17), i.e., of optimizing particular sectorial goals and strategies that neglect synergisms and trade-offs (see above, section 3.2) of cross-sectorial impacts of those strategies.

Coherence and consistency in policy formulation and implementation can be achieved by translating the different conceptual dimensions of and programmatic proposals for sustainable development into a commonly applicable policy framework. Such a framework should facilitate the specifi-cation of the role of each development programme in attaining the overall goal of sustainable development. The following proposes an outline of a planning and policy framework, based on the above-defined operational concepts of sustainable growth and development.

Framework structure

The incorporation of environmental concerns in development planning and policy formulation requires the introduction of space as an explicit dimension of an integrative framework. The spatial aspect is reflected in environmental science by the fundamental concept of the ecosystem. Ecosystems represent the area of interaction of a biotic community with its non-living environment (Odum 1971: 5). In development planning, local aspects have been addressed, in particular, in rural development and urban planning. In Table 4.1, three regional levels are identified that are particularly relevant for the organization of human activity. National boundaries are introduced to reflect the central role of governments in socio-economic planning and policies and in the integration of environmental concerns into plans and policies. Local aspects of the sustainability of development have been addressed in an 'ecodevelopment' approach to the integration of environmental and socio-economic activities at the grass roots of human communities and ecological systems. However, environmental impacts may spread beyond local ecosystems and national borders. Transboundary pollution, depletion and degradation of internationally shared natural resources and concern over the deterioration of 'global commons' such as climate change and the depletion of the ozone layer are examples of global environmental impacts. International organizations address those phenomena by means of commonly agreed strategies and conventions.

The framework also introduces four strategic functions of

1 assessment;
2 research and analysis;
3 planning and policies; and
4 support.

Those functions are commonly reflected in planning and policy-making at national and international levels (see, for example, in the case of an international action plan, United Nations 1973). The four functions are then confronted with the two basic concepts of sustainable economic growth and sustainable development at the local, national and international levels. The resulting framework matrix of Table 4.1 can thus provide information about particular strategy functions for all regional levels or, if read vertically, on comprehensive approaches to sustainable growth or development at national, sub-national and international levels.

Table 4.1 A framework for sustainable growth and development

	1 Local development		2 National development		3 International development	
Development paradigms	1.1 Regionally sustainable economic growth	1.2 Ecodevelopment	2.1 Sustainable economic growth	2.2 Sustainable development	3.1 Development of a supportive international economic environment	3.2 International and global sustainable development
Strategic functions						
A *Assessment*	Integrated regional (monetary) accounting	a Environment statistics b Regional (physical) resource accounting c Statistical ecology	Integrated environmental–economic accounting	a Environment statistics b Natural (physical) resource accounting c National development reports	International (comparative) environmental accounting	International (integrated) databases, statistical compendia and reporting
B *Research and analysis*	Modelling spatial disparities in income and growth	a Modelling ecodevelopment b Modelling carrying capacity c Development of ecotechniques	Integrated micro-, meso- and macroeconomic analysis and modelling	Models of integrated (physical) planning and development	International and global economic/environmental analysis and integrated modelling	International (physical) modelling of global environmental concerns and their socio-economic implications (for example, Global Input–Output Model)

C Planning and policies	Regional planning and policies of sustainable economic growth	Ecodevelopment planning and administration (decentralized planning and strategies)	a Reorientation of macroeconomic policies towards sustainability b Policies of structural change c Economic (dis)incentives for microeconomic planning and management	a Integrated (physical) planning and project formulation b Demographic policies on population, resources, environment and development c Programmes of human needs satisfaction	International strategies of sustainable economic growth (trade, investment, taxation, debt-for-sustainable-development swaps)	a International standards for sustainable development b International strategies and conventions (for example, Agenda 21, UNCED, climate, biodiversity)
D Support	a Extension service b Technical assistance (for local-level projects) c Public awareness building and participation (support to grass-roots movements/organizations)		a Public awareness building and participation b Programmes and projects of education, training and public information c Institution building and environmental law/regulations d Technical co-operation, including transfer (import) of environmentally sound technology and capacity building		a Promotion of sustainable growth and development at international and global levels b Multilateral support for international (global) sustainable development c International institution building (for example, United Nations Commission on Sustainable Development)	

Chapters 1 and 2 have already described key approaches to assessing sustainable development by means of integrated accounting and indicators of sustainable (human) development. Further aspects of data requirements for sustainable development will be discussed in Chapter 5 as a supportive means of implementing a national action plan. General analytical questions, including modelling of sustainable development, were raised in Chapter 3. Some analytical aspects will also be taken up below, when discussing the use of environmentally-modified variables in macroeconomic (national) policies and local-level modelling of ecodevelopment. The following chapters will, therefore, focus mostly on the central strategic function of planning and policies and related analysis.

Proactive vs. reactive policies

The purpose of the policy framework is to assist decision-makers in the formulation and implementation of consistent policies, strategies, programmes and projects, geared towards the common goal of sustainable development. Such action can be either proactive (anticipatory), addressing the socio-economic sources of environmental impacts, or reactive, dealing with environmental effects after their occurrence. The latter may comprise corrective (clean-up) or adaptive (for example, dykes to protect against sea-level rises) policies (Karshenas 1992: 19). 'Shifting the focus to the policy sources' of environmental impacts (WCED 1987: 310) reflects an anticipatory strategy that recognizes the key role of governments – both in causing environmental problems and in resolving them.

The anticipatory approach will have to adjust the conventional instruments of socio-economic policies and programmes. Such adjustment needs to be based on models that can predict environmental effects of socio-economic policies and programmes and that contain control (policy) variables permitting the avoidance or reduction of anticipated effects. In the reactive case, the environmental impacts will have to be assessed and clean-up action carried out by existing or newly created environmental institutions and protection activities. In both cases of environmental action, additional costs and benefits are generated. Their implications for long-term economic growth and short/medium-term changes in national income, employment or inflation are examined below.

Generally, the arguments favouring anticipatory policies rest upon the assumptions of

1 rising costs from accumulating environmental impacts;
2 foreseeable (with different degrees of uncertainty) prohibitive environmental effects (threat-to-survival argument); or
3 possible violations of social norms/standards such as intergenerational equity.

Reactive policies seem to be in order in cases of

1 clean-up of past environmental degradation;
2 high discounting of the future by individual or collective decision-makers; and
3 anticipation of technological breakthroughs for dealing with environmental impacts in the future, possibly prompted by anticipatory research (Pearce, Markandya and Barbier 1990: 7 *et seq.*).

Those policies need to be related to specific areas of environmental concern, such as global warming, pollution, deforestation, desertification or biological diversity, at different regional levels. For actual applications of the framework, for example, for drawing up a national action plan on sustainable development, a list of environmental concerns would have to be introduced as a further dimension for each of the framework's strategic functions. In this manner, a fully elaborated framework could provide for a systematic review and evaluation of available policy instruments, addressing the main environmental problems together with established socio-economic objectives.

4.2 Eco-variables in macroeconomics

Macroeconomic policies have traditionally focused on the achievement of overall objectives such as the 'magic square' of economic growth, full employment, balanced foreign trade and price stability. The objective of equity in the distribution of the results of economic performance is usually added as a supplementary goal, as it does not fit into conventional (economic) paradigms and models of general equilibrium and allocative optimality. The common wisdom of the discussion of environment and development since the Stockholm Conference (United Nations 1973) has been to request the addition of environmental objectives on a par with other economic and social ones in what could be called a 'magic polygon'. In the past, those requests were largely neglected by central powers since they came from the 'periphery' of weak governmental (environmental) agencies or non-governmental movements. More recently, political leaders have been facing increasingly outspoken and influential organizations and public opinion, bringing them to the conference table in the First World Summit Meeting on Environment and Development (UNCED) in Rio de Janeiro. Judging from their contributions to UNCED, they now appear to embrace the environmental cause as part of mainstream policies. Ways and means have to be found to harness any good intentions behind general statements and translate them into actual reformulation of socio-economic policies.

The above-described accounting identifies readily consumption, capital accumulation and foreign trade as the key variables in bringing about growth in income and product. They also reflect accounting relationships that

85

provide an inherent consistency in concepts and definitions for economic variables that have formed the basis of macroeconomic theory. Equation 2b of Box 2.4 indicates modifications of capital formation or investment by introducing the 'consumption' (depletion and degradation) of non-produced natural assets into a 'green' concept of 'capital accumulation' and by further introducing imports and exports of waste residuals across national borders.

Those relationships could be used as the starting-point for modelling environmental–economic interactions at the macro-level. Changes in the marginal propensities to consume, invest and import due to the introduction of environmental capital consumption and thus a reduced – potential – Net Domestic Product available for consumption, investment and exports could thus be explored in order to assess their effects on income and employment. Such modelling focuses on short-term stabilization policies for purposes of achieving macroeconomic equilibria in employment, trade, investment, saving, government budgets and foreign debt. Some of these short-term policy issues and their implications for environmental concerns are briefly

---

Box 4.2 Models of macroeconomic management for developing countries

Revised minimum standard model: the model was developed by the World Bank focusing on external debt. Based on a national accounts and balance of payments framework, it contains behavioural and technical relations among income, expenditure, investment, saving and the domestic and foreign credit markets. The model was tested and implemented in several island developing countries.

Computable general equilibrium models: the models are generally extensions of social accounting matrices in which behavioural equations are simulating the interactions of different economic agents (households, firms, government, etc.). Due to their complexity, they have been the focus of research rather than practical application.

Public sector planning and management information system: the system is conceived as a mixture of database, modelling and management system, consisting of different interlinked modules, *viz.*: the

1 policy analysis and planning system (for modelling at the macro-, meso- and project-level);
2 economic monitoring system (database);
3 financial management system (budget preparation and monitoring);
4 debt management system (debt service monitoring);
5 investment project bank (project monitoring); and
6 resource mobilization system (sources of project and programme funding).

Advanced computer technologies are expected to make this – hitherto untested – approach operational.

Source: Morrison (1993).

---

discussed below as they form the core economic policy in developed countries and – sometimes misguided (as to their socially disruptive effects) – structural adjustment prescriptions for developing countries.

Some promising extensions of basic accounting analysis for purposes of short-term macroeconomic management have been recently suggested (cf. Morrison 1993). They are described in Box 4.2. As these models are based on accounting frameworks, they seem to be readily adjustable to modifications of those frameworks as discussed in Chapter 2. Furthermore, the breakdown of environmentally adjusted variables by economic sector would facilitate the formulation of compatible policies of structural change in production and consumption patterns (see section 4.3, below).

Economic growth: consumption, investment, trade

As far as overall economic growth is concerned, it has been suggested that the availability of natural resource capital and the absorptive capacity of environmental media pose absolute limits on the level of economic growth. Consequently, economic policies would have to shift the focus from the 'efficiency' of resource allocation to the 'sufficiency' or limitation of the scale of economic growth. Sufficiency is reached at the optimal scale of economic growth where the 'long-run marginal costs of expansion are equal to the long-run marginal benefits of expansion' (Daly 1991c).

Uncertainties about the effects of technological progress, resource discovery and substitution possibilities, i.e., in the sustainability factors referred to in the above definitions of sustainability and sustainable economic growth, and problems of aggregating and predicting marginal cost and benefit functions cast doubts on any assessment of the closeness to such ultimate limits, both nationally and globally. As a consequence, recent international recommendations on the relationships between environment and development generally stress the need for continued or accelerated economic growth. For instance, the World Commission on Environment and Development concluded that 'the international community must speed up world growth while respecting the environmental constraints' (WCED 1987: 89). Box 4.3 presents a similar view from 40 'leading thinkers', convened to provide 'intellectual input' into the preparations for UNCED (Pronk and Haq 1992).

Sweeping statements on the desirability or necessity of economic growth, while protecting the environment, are political statements that can express a commitment to both objectives. The feasibility of attaining those objectives for concrete national, environmental and economic, conditions needs to be assessed, however, in terms of 'control' variables. Policy-control variables permit the application of available policy instruments and the monitoring of policy success or failure in terms of those variables. Replacing conventional growth indicators such as GDP or National Income by EDP

Box 4.3   Growth vs. environment

There is no choice between economic growth and environmental protection. It is an imperative. The issue is not *how much* economic growth but *what kind* of growth. Zero growth can be as detrimental to the environment as rapid growth. It is not the rate of growth that allows us to gauge the effect on the environment. It is the composition of GDP – the product mix as well as the types of production processes – that alone can tell us whether the overall impact on the environment is positive or negative. For developing countries, the choice is not to sustain poverty but to overcome this poverty through growth.

(Basic policy message 2, The Hague report (Pronk and Haq 1992: 10))

---

or ENI and expanding the scope of other relevant variables such as capital and capital formation to include natural capital and its use in models of economic growth can provide early-warning signals about the trends and limits of sustainable economic growth.

Taking long-term growth in EDP as the indicator of sustainable economic growth, three determinants of such growth can be readily identified from its definitory equation. They are consumption, investment (capital accumulation) and foreign trade. Their impact on the sustainability of economic growth, and policies reinforcing this sustainability are briefly reviewed in the following.

Changing consumption patterns   The role of the level and structure of consumption has received relatively little attention in conventional economic growth analysis and policy. At the international level, consumption patterns have been discussed in reference to non-sustainable lifestyles of industrialized countries and of the rich population strata in developing countries imitating those lifestyles. The global non-sustainability of Western consumption patterns is evoked vividly by Gandhi's famous question of 'how many planets will a country like India require if it took Britain half the resources of the planet to achieve its prosperity?' Clearly, the demand for consumption goods should share the responsibility of environmental depletion and degradation with the supply of those goods, irrespective of the immediate responsibility for particular environmental impacts. Environmentally sound production and consumption patterns are thus closely linked via the market interaction of supply and demand.

The role of market instruments in influencing production patterns through full-(environmental) cost pricing is reviewed below in section 4.3. As far as consumption patterns are concerned, changes in lifestyles away from wasteful and waste generating consumption have been advocated in particular in UNCED's Agenda 21 (United Nations 1993c: ch. 4). Those changes are to be achieved by reinforcing social – environmental – values through improved information (for example, environmental labelling of consumption goods) and education about the environmental consequences of consumption levels and patterns. Further research and analysis is required to determine the effects of changes in those patterns on overall economic growth and development, nationally and globally. Input–output analyses, based on integrated 'green' accounting tabulations, are particularly well equipped to determine the direct and indirect environmental costs of changes in consumption patterns – within and beyond the national economy. Box 4.4 indicates possibilities of transforming the production accounts of the SEEA into input–output matrices for purposes of input–output analysis.

Environmentally sound investment strategies   Capital net investment – over and above investment in capital replacement – is the engine of growth, i.e., the prerequisite for increases in national output. Development planning and analysis has in general focused on produced capital formation. For example, United Nations guidelines for different 'stages of development' of island developing countries recommend

1   investment in infrastructure and human resources for a 'big push' in the early stage;
2   investment in primary resources of agriculture, fisheries and tourism in the next stage; and
3   investment for the diversification of the economy and value-added generation in export industries thereafter (Corbin 1993a).

Box 4.4   Environmental accounting and input–output analysis

The SNA *Handbook of National Accounting – Integrated Environmental and Economic Accounting* (United Nations 1993a: ch. 5) describes also the transformation of its cost accounts into *input–output tabulations*. This is achieved by

1   reducing the information on accounts of produced and natural assets, focusing on capital accumulation only;
2   replacing the 'industry' classification by 'product' classifications; and
3   accounting for the environmental impacts of 'household production', i.e., extending the production boundary of the SNA.

Input–output analysis could thus examine, in particular,

1   the direct and indirect (antecedent) costs of environmental protection, i.e., the full amount of societal defensive expenditures (see above, section 2.2);
2   the international (cross-boundary) implications of flows of raw materials, produced goods and residuals/wastes (pollutants) in physical and monetary terms; and
3   the anticipated impacts of environmental protection on natural assets.

Human capital development has gained increased attention in the context of the new focus on human development, but also as the 'other' engine of growth. The latest International Development Strategy (United Nations 1991b) thus stresses particularly the importance of human capital for economic growth.

However, as discussed in section 3.1 above, economic growth depends not only on produced and human capital but also on the continued and possibly increased availability of natural and institutional capital. Due to the less tangible nature of their services, human and institutional capital hardly qualify for integration in monetary variables. By contrast, the introduction of produced and natural capital in integrated accounts enables investment policies to be based on a combined concept of net 'capital accumulation'.

Lack of adequate maintenance of produced capital is viewed in the above-mentioned guidelines as one of the most characteristic obstacles to development in poor countries. This key impediment to sustained growth has been largely neglected in development policies and strategies. Of course, this applies even more so to natural capital whose depletion, degradation, discovery or growth has been accounted for, to date, only in a few case-studies. The establishment of a trust fund whose yields would finance the (neglected) upkeep of produced capital has been proposed to deal with this maintenance gap (Corbin 1993b). Such a fund would be needed especially in the initial development stage, involving low-yield but high external economies generating infrastructural investment. Similarly, environmental funds for the maintenance of the natural resource base or its substitution by human or produced capital could be established (see section 5.3). The above-described (section 2.3) user-cost allowance anticipates such investment. If, however, investment opportunities are limited or the user-costing is not enforced, the generation of an 'exogenous' environmental (trust) fund with possible foreign aid support should be considered. As will be shown below in section 4.3, investment in natural capital maintenance can be considered as 'defensive', contributing to a 'structural distortion' of the economy. The total impact of such an effect could be assessed in models of input–output analysis (see Box 4.4, above).

Anticipatory policies would have to shift the focus of investment from restoration and rehabilitation of natural assets to investing in research and development of environmentally sound and resource-saving production processes. Such investment has been hampered in both the private and public sectors by difficulties of assessing the – time-lagged and external – benefits that frequently also extend beyond national boundaries (Droste and Dogsé 1991). Clearly, this is an area where far-sighted (in time and space) national and international policy-makers would have to allocate additional funds to facilitate such investment. Possible national and international arrangements in this regard are discussed below in Chapters 5 and 6.

Trade, natural resources and environment   Resource limitations inherent in the notion of sustainability can in principle be circumvented at national and sub-national levels by substituting nationally and locally wasting resources through trade (imports). This has happened in industrialized countries which now rely to a certain extent on the supply of energy and raw materials from developing countries. It is of course another question, as discussed above (section 1.2), to what extent there is enough room for substitution at the global level where the ultimate resource limits would be revealed.

On the export side, developing countries argue that improved access to international markets through trade liberalization would generally be beneficial to sustainable development. To quote UNCED's Agenda 21: 'An open, multilateral trading system makes possible a more efficient allocation and use of resources and thereby contributes to an increase in production and incomes and to lessening demands on the environment' (United Nations 1993c: para. 2.19). The argument behind this statement seems to be increased return from exports which could be invested into development. To the extent that such investment is environmentally sound and enabling the country to avoid further resource exploitation for the sake of survival (pollution-of-poverty argument), trade liberalization would indeed contribute to sustainable development in developing countries.

This argument needs of course to be examined under the concrete circumstances, i.e., production and consumption patterns, and investment policies and capacities of a particular country. Conceivably, improved market access might also lead to accelerated natural resource exploitation, the proceeds of which might be used up in additional consumption rather than investment in (man-made and/or human) capital. In this sense, trade and other domestic access barriers might very well have protective effects for the environment. Input–output analysis could be used to assess the direct and indirect effects of international trade in natural resources on national environmental assets, i.e., the amount of 'imported' or 'exported' environmental depletion or degradation (see Box 4.4, above).

Rather than relying on those somewhat ambiguous effects, the rational approach to integrated trade and environment policies is probably to use the benefits of the comparative advantages of trade liberalization, while, at the same time, fully accounting for the hidden externalities of trade in natural resources. Full-cost pricing for the depletion of natural assets and its environmental, social and cultural costs is the answer. If such costing exceeds the commodity prices, acceptable to large foreign corporations or the 'world market', exploitation of the respective resource would have to be halted.

Of course, such costing should be realistically assessed, for instance, through environmentally-modified national accounts. Where such accounting does not apply, due to significant intangible values attached to the environment, explicit conservation standards (for example, a ban on natural resource exports) would have to be contemplated. Indeed, a number of developing countries seem to have taken or are about to take such a step. In Thailand a nationwide logging ban has been established, and in Papua New Guinea (PNG) it was recommended that the costs and benefits of a log export ban should be assessed (World Bank 1990: ix). On the other hand, fears of pricing (taxing) out export industries might be often exaggerated. For example, a slump in mineral prices in PNG (in 1989) resulted in a negative 'net-price' (see above, section 2.4) which, under the assumptions of natural capital depreciation, would reflect pessimistic expectations about future cost and price developments. Nonetheless, it appears that mining operations are still flourishing in the country. Possibly, such a situation is the result of so-called transfer pricing which seems to be common practice, at least in the log market (see Box 4.5).

Full-cost pricing of exports is a special case of applying market incentives or disincentives in production and consumption; this is discussed in some detail in section 4.3 below. Obviously, different degrees of cost internalization affect the competitiveness of countries (or rather of domestic industries in international markets). From a 'liberal' point of view, suggestions to use trade regulation in multilateral agreements on the environment 'to tackle global environmental challenges' (United Nations 1993c: para. 2.20) seem, therefore, to be restrictive and prone to abuse (for trade rather than environmental purposes). In fact, any global regulation, be it through trade or direct international environmental conventions, which prevents the full

---

Box 4.5   Transfer pricing in PNG

A judicial Commission of Inquiry into Aspects of the Forest Industry identified the following practices of transfer pricing in the country:

1   misdeclaration of species;
2   downgrading of log grades;
3   underscaling of volume;
4   undervaluing log prices;
5   overvaluation of external management, machinery, lease and consultancy costs;
6   mistallying of ship loading (sinkers, lost logs, rejects, etc.);
7   overstated freight rates;
8   overstated claims for damages;
9   nil profit accounting;
10 third country invoicing (double invoicing);
11 complex distribution networks (importers' subsidiaries); and
12 collusion of importers.

Source: World Bank (1990: 37).

economic use of the natural wealth of a country would have to consider compensation for non- or reduced use if the economic benefits forgone exceed the share in environmental benefits from global provisions or regulations.

Many international conventions, such as the Basle Convention on the Control of Transboundary Movements of Hazardous Wastes and Their Disposal, the Convention on International Trade in Endangered Species of Wild Fauna and Flora or the Montreal Protocol on Substances that Deplete the Ozone Layer, affect international trade but do not contain compensation provisions. The complex and arduous international trade negotiations under the Uruguay Round seem to skirt this issue, and members of the General Agreement on Tariffs and Trade (GATT) seem to be anxious not to upset those negotiations by any amendments to the existing provisions of the GATT Article XX (see Box 4.6).

The provisions of Article XX refer only to domestic environmental phenomena. It is of course another question to what extent international environmental concerns should be addressed by trade sanctions as, for instance, had been the case on a US blocking of imports of tuna catch (from Mexico) that involved killings of dolphins (see *The Economist*, 27 February 1993: 25). Clearly, those questions need to be identified by GATT, especially if international environmental conventions contradict GATT stipulations. Some regional trade agreements appear to seek stronger environmental regulation and policy harmonization. The European Community's Maastricht Treaty and the North American Free Trade Agreement include environmental goals in order to reduce distortion in competitiveness from different national environmental taxation, subsidy and regulation.

The definition of EDP also refers to the possibility of involuntary 'trade' in residuals. As far as wastes and residuals are traded voluntarily, such trade

---

Box 4.6 Environmental exceptions to GATT obligations (Article XX)

The Article refers to measures invoking exceptions on environmental grounds as follows:

> Subject to the requirement that such measures are not applied in a manner which would constitute a means of arbitrary or unjustifiable discrimination between countries where the same conditions prevail, or a disguised restriction on international trade, nothing in this Agreement shall be construed to prevent the adoption or enforcement by any contracting party of measures:
> - necessary to protect human, animal or plant life or health
> - relating to the conservation of exhaustible natural resources if such measures are made effective in conjunction with domestic production and consumption.

can be considered as export or import of environmental (disposal) services
– already reflected in conventional trade balances. Illegal dumping or
pollution of other countries' territory, on the other hand, represents a free
service to the exporting country for which the importer should be compen-
sated. It is a matter of international agreement to what extent such exports
into territories outside national jurisdiction (the global commons of atmosphere or the oceans) should also require taxing nations for increases in the
deterioration of environmental quality they cause and compensating them
for health and welfare effects they bear. Accounting for international costs
'caused' and 'borne' (see section 2.2) would provide the data basis for the
application of these international economic instruments of cost internalization
(see also section 6.1, below).

Employment

Human capital can be viewed as just a further production factor employed
in combination with man-made and natural capital. Its efficiency (productivity) is determined to a great extent by the available production technology.
All these factors together determine the economy's capacity of supplying
goods and services and, if market clearance is obtained from the demand
side, balanced economic growth. In principle, the sustainability of growth
should thus be ensured by an allowance for the maintenance of human
capital in addition to those for produced and natural capital. As already
discussed (section 3.1), however, the employment of human capital represents a large portion of human activity with individual and social benefits
beyond productivity.

It is for this reason that full employment is typically advocated as a
separate, apart from economic growth, 'magic goal' in economic policy.
Thus, the discussion of trade-offs and complementarities between environmental and full-employment objectives has caught the particular attention
of the public media and the general public. The spotted-owl-conservation
vs. job-maintenance-in-logging-industries controversy in the USA illustrates
this widely publicized stance of environmentalists vs. economists. Of
course, a more rational discussion would have to resort to quantitative
analysis, comparing the (social) benefits gained from conservation of an
ecosystem and the economic benefits lost from production/employment
forgone.

A recent study of the International Labour Office (Doeleman 1992)
attempted to summarize national and international experiences with income
and employment effects of environmental policies and protection. Table 4.2
gives a synoptic view of this empirical compilation by country or country
groups. The study comes to the – tentative – conclusion that environmental
policy might generate additional employment, add to national production
and income in the initial stages of environmental protection but may

Table 4.2 Income and employment effects of environmental policy

Country	Environmental policy	Income and production	Employment
Austria	necessary environmental expenditure	−	
European Communities Countries	energy conservation and renewable energy use		+
Finland	water protection programme	+	+
France	pollution control expenditure	+	+
France, Germany, Italy, United Kingdom	SO_2 and NO_x scrubbers for acid rain control	+ (−)	+ (−)
Netherlands	doubling environmental protection expenditure	−	0
Norway	− carbon tax	−	0
	− environmental expenditure	+	+
Sahel	*no* action on land overuse	−	
United Kingdom	− pollution abatement expenditure	±	+
	− carbon emission tax (£ 30/t)	(−)	+
	− carbon emission reduced by 20%		+
USA	− environmental policy enacted	−	−
	− solar energy and energy conservation		+
Western Europe	non-specific environment policy		+

Source: Doeleman (1992: 44–50).

decrease income in the long run. The latter could be due to a crowding-out effect of fully employed factors of production by 'non-productive' environmental protection.

Such crowding-out would not have an immediate effect on conventional National Income and product (value added) to the extent it replaces final demand components (final consumption or capital formation). However, as discussed above, it has been argued that environmental protection does not contribute to an improvement in human welfare as it serves only to maintain the current status of environmental quality. Replaced final demand items, reflecting, for example, capital investment for environmental protection, could be considered as social 'defence' costs against deteriorating environmental conditions. Environmental expenditures would thus not generate an increase in 'true' national income and material welfare but may be responsible for a 'structural distortion' of the economy (Leipert 1984). Such an economy would tend to focus unwittingly on the maintenance of the quality of life rather than on its improvement. Structural change of the economy is proposed below for dealing with such distortion. On the other hand, in less-than-full-employment situations, job creation in any sector of the economy contributes not only to productivity but generates additional benefits of escaping the unemployed status and yielding job satisfaction.

Debt, finance, inflation, distribution

The following financial and distributional questions are interrelated, but their actual environmental implications are not always clear. The issues are raised here with a view to encouraging further research and assessment in concrete country situations.

Many developing countries experience a negative net transfer of financial resources resulting from private and public foreign debt servicing. Such transfer impedes the mobilization of domestic saving for investment and thus sustained growth. As far as the environment is concerned, it has been argued that the debt overhang might be an accelerating factor in resource exploitation. This is the case when 'mining' the resource presents the only means of fetching hard currency for debt-servicing. In some cases, structural adjustment programmes, which aimed at rescheduling debts or providing new loans, have resulted in 'adverse social and environmental effects' due to the need for budgetary reallocations (United Nations 1993c: para. 2.33).

Debt-for-nature swaps have been applied, though on a very limited scale, to address simultaneously issues of indebtedness and environmental conservation. They can be considered as a subsidy of foreign (private, non-governmental) organizations to prompt environmental protection in indebted developing countries. Box 4.7 outlines the mechanisms for swaps in two developing countries. Such swaps represent only one means of revenue raising from non-domestic sources. Their effectiveness has to be judged in comparison with other international support to national environmental policies (see Chapters 5 and 6 below).

Box 4.7   Debt-for-nature swaps

In 1987, the US-based NGO Conservation International bought an uncollectible debt of $650,000, owed by *Bolivia* to Citicorp, for $100,000. Bolivia's debt was forgiven in return for a promise to spend $250,000 in local currency for the protection of a buffer zone around a natural reserve.

In *Costa Rica*, some of the largest swaps were made. $11 million in grants and donations were used to purchase $70 million of debt and to exchange them through the Central Bank into local currency bonds of $36 million. Interest from the bonds is to support the country's natural parks.

Source: UNCTC (1991: T–20).

Transfers of financial resources from debt-servicing, grants or loans are already reflected in National Income. The combined effects of financial flows and environmental deterioration could therefore be assessed summarily by environmentally-adjusted National Income indicators such as the ENI. More comprehensive analyses would have to trace debt and credit flows and environmental costs through capital finance accounts for different sectors and institutions. The full incorporation of financial accounts and their distributionary effects into the proposed system of integrated accounting constitutes a further challenge to national accounting and accounting analysis.

Perhaps least explored are possible trade-offs between sustainable development and price stability. In principle, sustainable economic growth can be defined in real (price-deflated) terms, abstracting from inflationary processes in the analysis of such growth. There are, however, significant impacts on growth resulting from 'hedging against inflation', i.e., moving out from long-term investment into inflation-proof assets (land, gold, etc.) (Archibugi, Delors and Holland 1978). Another possible effect may result from additional environmental expenditures which could contribute to (cost-pushed) inflation in full-employment situations. Also, the focus of economic policies on short-term transactions in times of inflation could lead to a discounting of long-term social goals such as the preservation of natural assets for future generations. On the other hand, high discount rates could discourage investment and related demand for natural resources while at the same time accelerating the current exploitation of natural resources. Those ambiguities in potential environmental effects of changes in discount rates make the use of those rates for environmental policy and the achievement of intergenerational equity a rather ambiguous instrument (Pearce, Markandya and Barbier 1990: 151).

Social equity in the distribution of income and wealth is an elusive goal of economic policy. As an 'add-on' to otherwise well-integrated objectives of growth, employment, investment or trade, it is normative in character and subject to many non-economic determinants of fiscal and social policy. Despite a general recognition that, in particular in developing countries, the eradication of poverty is the central goal of development, anti-poverty strategies are therefore usually applied only after the implementation of economic – growth – policies. Full integration of distributional with environmental and economic concerns is thus probably better pursued through integrated planning and policies of broad-based sustainable development. As will be shown in Chapter 5, such development will have to pursue poverty eradication and equity by mutually reinforcing economic and demographic policies and strategies of human resource development, health and local-level development.

4.3 Cost internalization and structural change

Structural distortions

If indeed a large part of economic activity is devoted to dealing with environmental effects rather than to the generation of true (welfare-creating) income, a considerable distortion of the economic structure is the consequence of such policies (Leipert 1984). As discussed in section 2.2, integrated accounting alerts to such possible distortion by recording separately defensive environmental protection expenditures. Further structural distortions from malfunctioning markets and their – wrong – price signals to producers and consumer (Karshenas 1992: 22), and misguided governmental intervention in markets (Panayotou 1991) have contributed to environmental problems.

Four categories of structural distortion affecting or affected by environmental impacts can be identified:

1 deep-rooted structural distortion of underdevelopment caused by persistent poverty and population pressures;
2 price distortions due to under- or non-pricing of natural resources and environmental services;
3 failures of governmental intervention that subsidizes or otherwise prompts excessive uses of natural resources; and
4 an increasingly defensive structure of the economy in response to deteriorating quality of life and living conditions.

The last can be seen as a reaction to the former categories of distortion. Wrong valuation of environmental assets, considering those assets and their services as free, while in fact becoming increasingly scarce, is the cause of many environmental problems of overuse and destruction. The above-described environmental accounting approach reveals such wrong valuation by imputing the costs of depletion and degradation by private or public economic agents.

The first category of distortion points to a more fundamental cause of environmental disruption in developing countries where markets may have only a limited 'reach'. Pollution of poverty and related pressures of ever-increasing populations in both marginal rural areas and urban settlements have been identified above (section 1.2) as the driving forces behind environmental decline in developing countries. As these forces and their underlying social and political factors cannot be managed by market instruments alone, they need to be addressed by a broader 'development' approach as suggested above and further elaborated in Chapter 5 (as to their policy implications).

The following examines the use and usefulness of market-based solutions to structural distortions of resource allocation in the economy that are due to market malfunctions in dealing with environmental goods and services. The existence of such distortion is now widely recognized, even by industry. As one leading industrialist puts it: 'our markets are distorted by the underpricing of resources and services. And they suffer from policy failures such as subsidies' (Schmidheiny 1992). A recent World Bank report, for example, estimated that energy subsidies in former non-market (centrally planned) Eastern European economies accounted for half of their air pollution. Similarly, logging fees in five African countries covered only between 1 per cent and 33 per cent of the cost of replanting required for sustainable forest use (World Bank 1992: 11–12).

Internalization of environmental costs

The policy answer to this distortion of the economic structure is structural change of both the technological and sectoral structures of the economy (Jänicke et al. 1989). Such policy should not only redirect development from defensive towards welfare-generating activities, but should also encourage resource-saving and low-waste production and consumption. Public investment and support of private investment in environmentally-sound production and the encouragement of environmentally-friendly consumption patterns were discussed above (section 4.2). For developing countries, environmentally-sound technologies will have to be transferred 'on favourable terms, including concessional and preferential terms' (United Nations 1993c: ch. 34; see also section 5.3 below).

Overall policies need to be implemented at the level of actual decision-making. The strategic answer to structural distortion is thus to make the transition from accounting for imputed environmental costs to actual budgeting of those costs, i.e., to their internalization by economic agents. Open corporate accounting for actual and potential (probable) environmental liabilities has been proposed in a move from entrepreneurship to 'ecopreneurship' (Rubenstein 1989; see also Box 2.7 above). The next step should be – at least in the view of one entrepreneur – the achievement of eco-efficiency by means of minimizing natural resource use and pollution (Schmidheiny 1992).

In resource-intensive industries, such internalization should prompt exploration of more efficient resource use, including recycling and resource substitution or conservation. The objective is to approximate optimal levels of natural resource stocks and corresponding extraction rates through full-(depletion)cost pricing. Full costing does not necessarily imply non-use (preservation) of the resource which could be prodigal in the cases of sub-soil resources and even of 'non-critical' (higher-trophic-level) biological resources. The above-described user-cost allowance, for instance, suggests that even in the case of exhaustible (mineral) resources a 'permanent income stream' would continue to flow through reinvestment of the proceeds after the natural resource has been fully exploited (see Box 2.5).

The same and other industries, on the other hand, may become culprits in violating standards of environmental quality by producing pollutants (residuals) along with their desired outputs and through ad hoc pollution accidents and malpractices in natural resource and land use. Appropriate costing of these non-desirable joint products and side-effects at the sectoral level would alert to distorted (exaggerated) value-added generation by economic sectors and thus to distortions in the economic structure. In integrated accounting, such costing refers to the avoidance or mitigation of environmental impacts, depending on available technologies and/or 'exogenously' set standards and regulations. Of course, such costing of compliance with normative standards does not permit the determination of an optimal level of pollution, but rather a 'feasible' or 'acceptable' one, at any rate, a more practical one (Pearce, Markandya and Barbier 1990: 158; cf. also the discussion of sustainability vs. optimality criteria in section 3.3 above).

Policies of – sectoral – fiscal disincentives or incentives, which discourage investment and employment in environmentally-damaging industries or encourage alternative production and consumption patterns, address directly these distortions at the decision-making level. Such policies aim at the internalization of environmental costs into the budgets of enterprises and households. The basic rationales behind these (dis)incentives are the polluter-pays principle (including liability for environmental accidents and other damages) and the user-pays principle for the depletion of scarce natural resources. The aim in both cases is to make those who cause environmental problems accountable for their environmental impacts.

In environmental accounting, this incorporation would be reflected by turning imputed environmental costs into regular costs of capital consumption or intermediate consumption. In this manner, market prices formed by individual supply–demand interactions would tend to reflect all long-term marginal or average costs. Those costs would include not only direct environmental damage and depletion costs but also outlays required for the avoidance of such cost, for example, through recycling or the introduction of alternative production patterns and technology. Other (social) costs of development such as those of achieving current and intergenerational equity could also be included in principle in full-cost pricing. Ultimately these costs would be shared between consumers and producers according to prevailing price elasticities of supply and demand (Bartelmus 1992a).

Another argument for the internalization of those and, in fact, any kind of social costs are the optimization properties of general equilibrium models. Those models assume perfect competition, ensuring that prices equal marginal costs, including internalized costs, thus bringing about optimal resource allocation. This is of course standard microeconomic textbook knowledge (see in the environmental context, for example, Pearce, Markandya and Barbier 1990). The question is if perfect conditions do not prevail in reality, then to what extent is the application of 'economic instruments' for social cost internalization still justified. Actual experiences with these instruments are discussed briefly below.

Economic instruments

Economic instruments – as opposed to regulatory ('command-and-control') measures have met with increased interest as a means of implementing the polluter- and user-pays principles. The recent preference for economic instruments can be traced back – according to an OECD (1989) survey – to their conformity with several policy trends, *viz.*:

1 a tendency towards reduced government intervention;
2 the search for more cost-effective solutions to environmental problems;
3 the need for policy integration due to interdependences between social, economic and environmental activities and effects; and
4 the transition from reactive to proactive (preventive) environmental policies.

The current widespread rejection of central economic planning in Eastern European countries and its 'newly discovered' environmental and economic inefficiencies strengthen the argument for applying market-oriented instruments to encourage environmentally-sound and resource-efficient production and consumption behaviour. The efficiency of market (dis)incentives lies in achieving full costing at the microeconomic level, thus permitting individual economic agents to choose among a range of alternatives in matching costs with benefits (income, profit). Assuming rational behaviour of those agents, they can be expected to find least-cost approaches (at given market prices) under different economic and technological conditions and options. Central command/control measures will typically lack the information, capacity and motivation to identify and enforce those options for different agents. It is indeed this argument which makes for a reduction of government intervention in markets as indicated in point 1 above.

The advantages of preventive action and need for reactive responses were discussed above in section 4.1. Most economic instruments attempt to bring about behavioural changes that address environmental problems before their actual occurrence. Such an anticipatory stance is reflected explicitly in a proposal to create environmental assurance bonds which would account for possible future environmental impacts and resulting liabilities. Those bonds would either be used to meet actual liabilities or would be paid back if no impacts occurred. Assurance bonding could thus be considered as 'an extension of the "polluter pays" principle to "the polluter pays for uncertainty as well"' (Costanza 1991: 77).

Box 4.8 lists the main instruments of providing economic (dis)incentives to economic agents for the internalization of environmental (social) costs. There are, however, limits to the useful application of economic instruments.

Command-and-control measures would have to be applied where acute, high-risk environmental effects require immediate action rather than time-lagged prompting through incentives. Life-threatening substances in the food chain or land contamination with toxic wastes are examples of such high-risk phenomena. A further case for non-market regulation is typically made when individual preferences and decisions are considered to be short-sighted with regard to social goals whose desirability should be established by the far-sighted vision of experts or the government. Such short-sightedness, or disagreement with debatable value-laden objectives, might effectively prevent implementation and enforcement of fiscal measures. This is the concern of 'deep ecologists' (see above, section 3.3) who attach existence values to environmental assets, for example, to particular species, calling for their conservation by governmental fiat, through standard setting, laws or regulation, and enforcement. Similarly, distributionary aspects, especially of market incentives, need to be carefully assessed, since they might offset any gains in environmental benefits of protective action.

More practical obstacles also need to be considered. Problems of measuring accurately synergistic and cumulative environmental effects from multiple sources in particular areas, and, in turn, attributing overall effects to causing agents make it quite impossible to determine the correct marginal social environmental cost and corresponding charge, i.e., the so-called Pigovian tax (Zylicz 1991: 386). The costing of compliance with national environ-

---

Box 4.8 Instruments of environmental cost internalization

Most economic incentives or disincentives fall into the following categories:

1 *Taxes/charges* on pollutants or potentially polluting or 'wasting' products, such as effluent charges on the quantity of different emissions, fees on natural resources exploitation (beyond regenerative capacities), or negative taxation (*subsidies*), for example, grants, loans or tax exemption for environmental protection, as well as the removal of subsidies for environmentally unsound production.

2 *Deposit-refund systems* which surcharge potentially polluting production: when pollution is avoided, the surcharge is refunded, applying a similar principle as in the above-described pollution assurance bonding.

3 *Generation of markets* by means of establishing individual 'property rights' to the use of previously 'commonly' (governmentally) owned natural resources, including rights of using those resources as a sink for pollutants, for example, tradable pollution permits.

4 Less tangible *prompting* of private households (but more and more also corporations) to change environmentally unsound consumption (and production) patterns: typically such prompting is conducted through environmental campaigns or through the educational system to obtain 'green consumerism' (Pearce, Markandya and Barbier 1990: 155).

102

mental norms and standards is a pragmatic solution which, however, does not meet with general (equilibrium) optimization criteria. On the other hand, such costing would provide a direct link between criteria of sustainable development at the macro-level (as mentioned above) and the use of economic instruments for bringing about environmentally sound behaviour of economic agents at the micro-level. High administrative costs of monitoring and control are further impediments to making economic (dis)incentives the main or only approach to environmental policies.

In developing countries, the 'limited reach' of fiscal policies might be the main obstacle to applying environmental taxation (Desai 1991: 19). A good, but perhaps extreme, example is the situation in Papua New Guinea where over 90 per cent of the land is owned in customary tenure by clans and tribes. A recent seminar on the implementation of UNCED recommendations in the country (see section 5.3) revealed that governmental influence on modifying environmental behaviour at the local level is largely restricted to awareness-building. Measures of market incentives and regulation can thus be expected to fail in most cases.

Other obstacles to the application of economic instruments in developing countries are

1 the relative low priority (as compared to economic development) given to environmental issues;
2 distributionary effects that might burden the poor and prevent them pursuing their already limited economic activities; and
3 the lack of institutions for monitoring environmental effects, compliance with standards and enforcing the application of economic instruments.

Rather than introducing new fiscal instruments in developing countries, the removal of subsidies that encourage resource depletion (for example, the colonization of the rain forests) and of the use of polluting products (for example, fertilizers and pesticides) has therefore been suggested 'as an initial step' (Doeleman 1992: 92, 95). The removal of such subsidies addresses directly the above-mentioned structural distortion of markets by misguided governmental interventions.

Most industrialized countries already apply environmental charges. Table 4.3 shows the application of different types of charges in OECD member countries. One assessment of the usefulness of market incentives for environmental policy in those countries found that these tools were quite successful in revenue raising but failed in changing individual behaviour. The main reason seems to be a too timid setting of charge or tax rates (OECD 1989). Another reason might be the existence of monopolistic or oligopolistic markets which permit the transfer of environmental costs to other parties. The above-described measurement of environmental cost in integrated accounts should facilitate a more realistic assessment of the level of charges than the current trial and error approaches. They would also

Table 4.3 Application of effluent charges in OECD member countries

Country	Effluent charge Air	Water	Waste	Noise
Australia		x	x	
Belgium			x	
Canada				
Denmark				
Finland				
France	x	x		x
Germany		x		x
Italy		x		
Japan	x			x
Netherlands		x	x	x
Norway				
Sweden				
Switzerland				x
United Kingdom				x
United States			x	x

Source: OECD (1989: 34).

provide the basic data for the further analysis of the cost ultimately borne by economic agents according to prevailing market (in)elasticities.

The problem with increasing charges is, however, that there is a general resistance to any tax increase which has, for instance, effectively thwarted the introduction of higher environmental taxes in France and Italy (OECD 1989: 114). Such resistance has also been observed in industries that are prone to regulation. For instance, the utilities industry in the USA seems to shy away from purchasing emission permits. Obviously, they prefer complying with regulations to trading risky pollution permits (*New York Times*, 25 January 1993: D2).

In the USA an attempt has been made to overcome such resistance with the argument of 'taxing the bads' rather than the 'goods of income and labor' (Dower and Zimmerman 1992: 8). Studies of the World Resources Institute claim that green charges and levies would indeed 'allow much more burdensome taxes to be reduced by 10 percent or more' (Repetto *et al.* 1992: 1). Box 4.9 shows widely differing estimates of the levels of a carbon tax required to reduce CO_2 emissions; at the same time, a relatively narrow (and low) range of effects on GNP is indicated.

In line with the discussion of the limits of monetary valuation in integrated accounting and corresponding economic policies and taking into account the above-described drawbacks of economic instruments, alternative or complementary approaches need to be developed. The following chapter will explore possibilities of integrative (development) planning and decision-making – both at the national level, and at the grass roots of traditionally owned and managed natural resources.

Box 4.9   Features and economic effects of a carbon tax in the USA

Scientists of the World Resources Institute have proposed a carbon tax to reduce CO_2 emissions in the USA and as a contribution to lowering global greenhouse gas emissions. The tax is considered to be cost-effective and generating revenues which can be used to offset negative distribution effects; it may even result in net gains for the country.

The carbon tax is an excise tax on the producers of fossil fuels, based for practical reasons on the carbon content of those fuels. It is thus not an emission charge. However, carbon content is considered to be proportional to CO_2 emission, once the fuel is burnt.

Estimates of the macroeconomic cost of carbon reduction to 80 per cent of the 1990 level by the year 2020 range between baseline GNP decreases of 1.6 and 3 per cent if revenues are not 'recycled', i.e., used to reduce the tax rate on capital or labour. If such recycling would be carried out the range is between an increase (gain) of 1 per cent and a decrease of GNP of -2.2 per cent. The differences in those estimates can also be explained by widely differing estimates of the carbon tax rate required to achieve the same goal of carbon reduction: carbon tax rates proposed range between \$40 and \$250 per tonne of carbon. It should be noted, however, that those costs and gains do not take the – difficult-to-measure – environmental benefits into account. The realization of those benefits are after all the main objective of environmental taxation.

Source: Dower and Zimmerman (1992).

5

PLANNING AND POLICIES II: SUSTAINABLE DEVELOPMENT

The present chapter continues the discussion of the planning and policies function in the policy framework introduced in section 4.1. Chapter 4 dealt with the modification of economic policies to achieve sustainability of economic performance and growth. The limited reach of those policies was demonstrated, referring to development objectives beyond the sustainability of produced and natural capital, and to local-level activities largely outside the monetary market system. These two aspects of development are the topics of the first two sections of the present chapter. The last section discusses some means and two case-studies of implementing a green development or action plan.

5.1 Integrated planning and policies

Integrated development planning

Economic policies are typically integrative, focusing on variables that use the *numéraire* of the monetary unit for aggregation and disaggregation across sectors and institutions. As shown above, certain aspects of sustainability lend themselves to such monetary aggregation but find their limits with increasing distance of development objectives from the economic process, or where social concern overrides individual valuation implicit in market price formation. Development goals of a social, especially distributionary, cultural or political nature are thus hardly possible to value in monetary terms. The policy focus on monolithic monetary measures of economic growth has therefore been criticized by advocates of multi-objective development that aims at the improvement of the human quality of life (see section 3.1 above).

The more radical concept of the steady-state economy which develops without growing (Daly 1991c) enlarges on these criticisms, assuming that limits to 'physical' economic growth have been or are about to be reached. The only possibility for development is to realize 'qualitative potentialities' or, in the present terminology, non-economic development objectives.

The establishment of an explicit normative framework of standards and targets has been suggested above for the integration of these objectives in comprehensive multidimensional development analysis.

Comprehensive integration of non-monetary variables has not been commonly applied in development planning and policies, due to conceptual as well as statistical problems. First of all, there has been widespread scepticism about the efficacy of development planning. 'Divorce between what is planned and what actually takes place is rather a common feature among developing countries' (Abdalla 1977: 157). There is no reason to believe that this blunt statement, made 17 years ago, is not valid any more. Box 5.1 lists some of the more pertinent reasons for ivory-tower plans which never had a chance to be implemented the way they were designed.

An integrated programming approach to sustainable development, determining activity levels within a feasibility space, defined by capacity and other maximum and minimum constraints, was described above (Box 3.3). In practice, however, the use of models of integrated development planning has not been very successful at the national level. The main reasons are

1 difficulties of central planning in terms of large sets of mostly non-comparable physical variables; and
2 the societal, in particular governmental, preoccupation with short-term economic fluctuations, as compared to long-term (beyond legislative periods) goals of environmental resource conservation and intergenerational equity.

There are indications, however, that this pessimistic view of development planning is changing. National and international development planners and

---

Box 5.1 Ivory-tower plans: the reasons for plan failures

1   overambitious plan objectives;
2   ignorance or neglect of conflicts between plan objectives;
3   insufficiency of the database for plan design, monitoring and evaluation;
4   lack of qualified personnel to design and carry out plans;
5   inaccurate and inconsistent plans (due to data gaps, neglect of conflicting objectives and a lack of qualified planners);
6   lack of communication between planning agencies, day-to-day decision-makers and those affected by decision-making;
7   unexpected events such as sudden deterioration of terms of trade, natural disasters or political upheaval;
8   corruption;
9   lack of political will to carry out the plan.

Sources: Todaro (1977: 376–8); Islam and Henault (1979: 257); Seers (1983: 91–129); UNDP (1991: 6).

---

authorities appear now to be taking a renewed interest in an integrated development approach that includes non-economic, demographic, social and environmental variables. Section 5.3 describes two case-studies of integrative planning, established in anticipation of, and response to, UNCED. It remains to be seen if these two examples represent indeed a general revitalization of broad-based development planning. It is interesting to note, on the other hand, that international deliberations seem to avoid concrete references to a development plan. Rather, oblique references to 'planning', 'policies' or 'strategies' are typically made, avoiding, in fact, any critical assessment of the success or failure of the planning process. This holds for the global International Development Strategy for the Fourth United Nations Decade (General Assembly resolution 45/199), as well as the discussion of sustainable development by UNCED.

The following reviews broader approaches to sustainable development in two stages. First, the extension of sustainable economic growth policies by including human capital and the improvement of its productivity through human resource development is discussed. Then, further policies and strategies for meeting non-economic goals of 'human development', notably the eradication of poverty, are reviewed. In particular, the role of demographic policies in both poverty alleviation and environmental sustainability is examined.

Neglected productivity: human resource development

The effects of environmental protection on employment were discussed briefly in the preceding chapter (section 4.2). Probably more important is the reverse effect of human capital on development as the 'other' engine of economic growth (in addition to produced capital). This role was somewhat neglected in the past. This is not surprising since national accounts do not include human capital as an asset on a par with produced capital. The reason is obvious: human 'assets' (beings) are not owned and traded and thus valued in the markets as is the case for produced assets. Investment in human capital is thus not recorded as capital formation, and therefore has a relatively low visibility in categories of consumption expenditures for education or health.

This view of human resources seems to have changed. In the face of widespread failures of national and international development strategies, human resource development has now become the latest battle-cry of growth-based development. The Fourth International Development Strategy of the United Nations (1991b) thus proclaims 'human resource and institutional development' as one of four 'priority aspects' of development (the others being eradication of poverty and hunger, population, and environment). The strategy also stresses the close interaction of human resource development with the process of economic and technological transformation. The former, i.e., the formation and adaptation of labour supply to overall, sectoral and regional demand is briefly discussed in the following; the latter, i.e., the use of new and environmentally sound technologies, is described in section 5.3 as one of the 'means' of implementing sustainable development.

Improving the quality of human capital through health, information and education policies and programmes has immediate effects on the sustainability of economic growth – provided human capital is a limiting factor of production. Such limitation includes not only the quantitative supply of labour but also the skills required for particular production processes and technologies. Recent United Nations guidelines on integrated macro-economic planning and management distinguish between four components of human resource planning (Bartsch 1993):

1 population planning – responsible for planning for the potential labour force;
2 employment planning – balancing aggregate labour supply and demand;
3 manpower planning – balancing supply and demand by occupation and skill; and
4 educational planning – providing education and training for expected job opportunities.

Population planning and policies are of course motivated not only by economic (labour market) considerations. Social, cultural and environmental considerations are also significant determinants of demographic policies and are therefore discussed in the following section in a broader – development – context.

Employment planning at the aggregate level should be integrated by introducing labour and, if at all possible, the maintenance (sustainability) of human capital as explicit variables into models of sustainable economic performance and growth. Manpower planning and policies largely make use of incentives and disincentives of a financial (salaries/wage structure and other benefits) and non-financial (working conditions, career development, social status, vocational training, etc.) nature. They are to attract manpower to sectors, regions and occupations where required, i.e., where demand for skills is not met by supply.

Those incentives can of course also be used for encouraging environmentally sound production patterns, supporting in particular the application of new and environmentally sound technologies. This requires compatibility of incentives for reducing labour market distortions with the above-described instruments of environmental cost internalization. Similarly, educational planning should not only address potential imbalances of the labour market but should be harmonized with policies of education and information on environmentally sound consumption patterns. Education and training are considered below as a prerequisite for the successful implementation of sustainable development policies at national and local levels (see section 5.3).

As already indicated, the health aspect of human capital is difficult to assess. Clearly, health care, the prevention of chronic diseases, the provision of sanitation, safe drinking-water, nutrition and environmental protection contribute to the productivity of human capital. However, health and related policies also affect directly human well-being and are thus an important 'end' of human development.

Beyond productivity: human development and poverty eradication

Analytical approaches to defining and modelling sustainable development or, what appears to be the same, human development have been discussed extensively above (in section 3.2). In its latest *Human Development Report*, UNDP (1992b: 24) recognizes the need to translate ideas into practical action. Such translation is to be achieved through a new framework for national planning, new techniques for project appraisal and new strategies for aid. However, there is very little elaboration of the means of implementing human development in the report. Section 5.3 will further discuss those means – and additional ones of developing integrated databases and institutional capacities.

In the following, the focus is on the social dimension of sustainable development, i.e., 'human' development in a more narrow sense. This dimension manifests itself most conspicuously in developing countries in extreme conditions of poverty. The interactions between poverty and environment have already been described (in section 1.2) in terms of 'pollution of poverty'. Here, some thought is given to the policies of poverty eradication and, more generally, to population policies and their role in achieving or at least supporting sustainable development.

UNCED's Agenda 21 devotes one chapter to 'combating poverty'. It introduces the concept of sustainable livelihood as an 'integrating factor that allows policies to address issues of development, sustainable resource management and poverty eradication simultaneously' (United Nations 1993c: para. 3.4). An effective strategy for enabling people to earn a sustainable livelihood should cover a wide range of issues, including:

1 demographic factors;
2 income generation and distribution among social groups (focusing on disadvantaged groups of women, children, youth and refugees) and local groups (communities);
3 investment in human capital; and
4 rehabilitation of degraded resources and sustainable use of natural resources for basic human needs (food security, health care, access to water and sanitation).

Policies of investing in human capital (resources) development were discussed in the previous section; they aim at generating employment, income and economic growth. If those policies are targeted at disadvantaged social groups and local communities, they contribute directly to an improved distribution of income. In this manner, they can supplement or reinforce fiscal measures of income and wealth distribution. Possible synergisms between the redistribution of income/wealth and environmental protection could result under conditions of pollution of poverty or affluence (see above, section 1.2).

There is still little experience with the relevance of such synergistic effects for the combined pursuit of equity and environmental protection. Nor is it clear how criteria of intergenerational equity, as expressed, for instance, in the WCED's definition of sustainable development (see section 3.2), could be built into the set of development constraints described above for feasible development. It is of course a political decision to specify those criteria or targets, especially with regard to possible trade-offs for the well-being of the present generation. To the extent that pursuing a sustainable livelihood involves the maintenance of natural assets, those assets would be preserved for the use by future generations as well, introducing an element of intergenerational equity.

The following will briefly dwell on the remaining issues of achieving a sustainable livelihood, *viz.*, first, the role of demographic policies in poverty alleviation and environmental protection and, second, measures of meeting key human needs of health, shelter and food, and their interaction with environment and development planning and policies.

Population policies The role and use of population policies in sustainable development is controversial. It depends on the interaction of demographic factors of population growth, migration and distribution (concentration) with the generation or reduction of poverty and the effects and repercussions on/from the environment. Some correlational analyses seem to confirm a linkage between levels and growth of populations and environmental depletion and degradation, as well as poverty (Dasgupta 1992; Myers 1992). International proclamations appear to accept those analyses as a justification for the combination of population concerns with environmental concerns into a 'holistic view of development' (United Nations 1993c: para. 5.16).

The United Nations Population Fund (UNFPA 1991: 5) is perhaps the most outspoken protagonist of those views, blaming population growth for two-thirds of the increase of carbon dioxide emissions, 80 per cent of tropical forest depletion, the dwindling and degrading of freshwater resources and the degradation of coastal areas. The above-discussed (section 1.2) argument of having reached the limits to photosynthetic conversion by human activity, i.e., the limits of consumption of renewable energy sources

111

by biota, has been used to justify intensified efforts of reducing birth rates through family planning in order to avoid an 'ecological catastrophe' (Rasmuson and Zetterström 1992: 74).

The view that family-planning efforts have only a marginal influence on fertility with pay-offs only in maternal health and related social areas (Ohlin 1992) contradicts directly this argument. The same author holds that population growth in developing countries is a transitory phenomenon, comparable to demographic trends observed in the development of – now – industrialized countries. Moreover, there seems to be 'no evidence that a reduction in population growth will actively *reduce* existing levels of poverty' (Rodgers 1984; cited in Shaw 1989). Following these lines of argument, overall population policies should follow their own demographic and social goals, but should not be considered as an instrument of poverty eradication.

Similarly, the correlation argument regarding environmental impacts and population growth can be seen as an example of misinterpreting 'correlation' that – as every statistician knows – does not necessarily reflect cause–effect relationships. Thus, the popular view that demographic pressures are responsible for exceeding the carrying capacity of national territories or the planet has been explained as a superficial view of the 'tip of the iceberg where large, congested populations and environmental degradation are most visible' (Shaw 1989: 202). More significant determinants of environmental effects are hidden below the water, including

1 the impacts of technology;
2 conflicting interests of landowners and the landless regarding land and natural resource use;
3 marginalization of weaker groups into areas of low carrying capacity;
4 public mismanagement of natural resources; and
5 warfare (Ohlin 1992).

In conclusion, overall population policy seems to be hardly effective in tackling poverty and environmental problems. This statement has to be qualified, however. Population pressures on the carrying capacity of lands are typically unevenly distributed and depend on the nature (resilience) of the particular ecosystems affected. High population densities in vulnerable areas have indeed been shown to affect environmental quality and resource availability (see, for example, a review of population effects on land-use in Latin American and Sub-Saharan countries by Bilsborrow and Okoth-Ogendo 1992).

It is not surprising, therefore, that the discussion of Population, Resources, Environment and Development (PRED) typically descends to local levels once an attempt is made to translate integrated approaches or frameworks for PRED into concrete programmes or project proposals. An example of this descent is the contribution of the UNFPA (1991) to UNCED which,

112

after exhorting the need for policy integration and ministerial co-operation, immediately advances to the discussion of institutional and programme improvements at the village level. Similarly, key elements of UNCED's programmes for combating poverty, and for population and sustainability, are the empowerment of communities and the implementation of integrated environment and development programmes at the local level (United Nations 1993c: chs 3 and 5.C). Section 5.2 will, therefore, examine the main features of achieving sustainable development at the grass roots, in a local-level 'ecodevelopment' approach.

Basic needs satisfaction: health, food and shelter strategies The above three basic needs categories have one thing in common: their satisfaction is conspicuously absent in situations of poverty. It is impossible to describe here all the particular strategies and technical solutions addressing those conditions of poverty. A few observations on how key strategies can be linked to the overall planning process could be, however, the starting-point for a more coherent approach to achieving a 'sustainable livelihood for all'.

The dependence of human health on a broad spectrum of physical, spiritual, biological and economic/social factors has been recognized by the World Health Organization and confirmed by UNCED. Agenda 21 (United Nations 1993c: ch. 6) thus proposes a wide range of programmes, addressing primary health care needs, communicable diseases, vulnerable groups, urban health and health risks from environmental pollution and hazards.

As discussed above (section 3.2), one way to integrate health programmes with other sustainable development programmes would be to specify quantifiable targets and to assess their compatibility with socio-economic and environmental capacities and standards. Such targets were established for the programme on the control of communicable diseases as shown in Box 5.2. Those goals were recommended in international forums for national implementation. The box shows, however, that even in the relatively concrete area of disease control, no such targets could be agreed upon for AIDS and tuberculosis.

The objective of sustainable agriculture and rural development advanced by UNCED (United Nations 1993c: ch. 14) is to increase food production in a sustainable way and enhance food security. As pointed out by the WCED, the latter is not only a question of raising food production but of shifting the focus of production to food-deficit countries. However, current strategies of food production by governments

1 lack an ecological orientation (encouraging environmentally unsound farm practices);
2 lack a regional policy differentiation; and
3 use wrong incentives, subsidizing growth of production at the expense of conservation (WCED 1987: 132, 133).

Box 5.2   Selected goals for communicable disease control

1 by the year 2000, to eradicate guinea worm disease and polio;
2 by the year 2000, to effectively control river blindness and leprosy;
3 by the year 1995, reduce measles deaths by 95 per cent;
4 by the year 2000, reduce the number of deaths from childhood diarrhoea in developing countries by 50–70 per cent;
5 by the year 2000, to maintain the transmission-free status of areas freed from endemic malaria;
6 by the year 2000, to reduce the prevalence of schistosomiasis and of other trematode infections by 40 and 25 per cent, respectively, from a 1984 baseline;
7 to mobilize and unify national and international efforts against AIDS;
8 to contain the resurgence of tuberculosis.

Source: United Nations (1993: para. 6.12).

Local-level approaches of integrated regional planning and participatory 'ecodevelopment' address those concerns to some extent (see section 5.2, below).

Human settlements, especially urban settlements, generate environmental problems but can also contribute to their solution. However, 'little consideration' has been given to the human settlements aspects of sustainable development; this is despite the fact that 'settlement systems are the framework within which economic activities operate' and that they provide 'the spatial context for most capital-investment decisions' (UNCHS, undated: 3). Human settlement planning and strategies could thus provide the links between the participatory local-level ecodevelopment approaches, discussed below, and the top-down macroeconomic growth/investment policies, discussed above. Such interactive approaches are largely non-existent, as 'planning in most local governments is a rudimentary form of land-use zoning, usually unrelated to any socio-economic framework' (*ibid.*: 11). Box 5.3 lists a set of programmes that has been considered to improve the social, economic and environmental quality of settlements and their working environments.

Even this brief review of three categories of human needs already indicates numerous interdependences, not only between the overall macro-policy framework and local-level approaches to integrated development, but also among strategies of human needs satisfaction themselves. For instance, the provision of safe food, drinking-water, sanitation and shelter in human settlements is a preventive means for reducing the incidence of water-, food-borne and communicable diseases. The pay-offs in terms of human health and well-being of coherent integrated and preventive planning – beyond, but not exclusive of – sustainable growth policies are evident. Such considera-

114

Box 5.3   UNCED programmes for sustainable human settlements
development

1   provision of adequate shelter;
2   urban settlement management;
3   sustainable land-use planning and management;
4   integrated provision of environmental infrastructure of water, sanitation,
    drainage and solid-waste management;
5   sustainable energy and transport systems in human settlements;
6   human settlements planning and management in disaster-prone areas;
7   enhancement of the construction sector;
8   human resource development and capacity building for human settlements
    development.

Source: United Nations (1993c: ch. 7).

tion may indeed bring back the basic human needs approach to development
(see section 1.1) – with a twist, however. The twist would be the explicit
linkage to economic growth, i.e., to achieve what has been termed socially
necessary (Jayawardena 1991: 8) or 'desirable' (Karshenas 1992: 24) economic growth. Such growth is to ensure the satisfaction of basic human
needs as a minimum (see also section 6.1, below).

5.2 Grass-roots strategies: ecodevelopment

Despite the new interest in comprehensive development strategies and
indices, integrated modelling and planning might still have better chances
of implementation at the local level. Many interactions between human
activities and the environment are best observed and managed *in situ*, i.e.,
by those directly affected, rather than by planners and policy-makers in
distant capitals or administrative centres. This applies particularly to the
agrarian societies of developing countries. Their social and economic
systems are closely tied to the rhythm and productivity of nature's supply
of 'primary' resources of agricultural land, pasture, water and fuelwood, i.e.,
to the carrying capacity of bioproductive systems. It is for this reason that
local approaches to the implementation of environmentally sound and
sustainable programmes and projects are explicitly referred to in Table 4.1
under the generic label of ecodevelopment.

The needs and capabilities of indigenous people and other local communities have been recognized by UNCED in its Rio Declaration (principle 22)
and throughout the Agenda 21. Governments are requested to support these
communities within 'a new framework that integrates demographic trends
and factors with such factors as ecosystem health, technology and human
settlements, and with socio-economic structures and access to resources'

(United Nations 1993: para 5.42). 'Empowerment', 'involvement' and 'participation' are the buzzwords for enabling local communities to achieve sustainable development as perceived by themselves.

Those are indeed the very same criteria advanced for local-level integration of environment and development from the bottom up, i.e., ecodevelopment, nearly two decades ago (UNEP 1975). However, after a flurry of widely publicized case-studies of mixed success, the term ecodevelopment seems to have disappeared from the vocabulary of local-level approaches to environment and development. One reason might be the variety of particular environmental, social, cultural and political conditions that make each effort to implement ecodevelopment different and thus discourage generalization. Nonetheless, as described above, the need to deal with issues of poverty, demography and vulnerable groups at the local level suggests the revival of ecodevelopment – if only to keep track of earlier experiences with this approach and to avoid some of the fallacies encountered.

The following review of the interaction of humans with ecological systems is to provide the generic framework from which some general criteria for ecodevelopment planning and implementation can be derived. As already indicated, it is difficult to go much beyond such generalities. Rather, the discussion of some case-studies and their underlying models and frameworks might shed light on how to initiate ecodevelopment programmes and projects. In a next step, the particularities of the regions and its inhabitants, i.e., their needs and aspirations, would have to be built into the generic approach. The section concludes with some discussion of the links and conflicts between bottom-up ecodevelopment and top-down 'regional planning'.

Man and ecosystem

The concept of the ecosystem has been used to model the interactions of populations with the non-living components of their environment in a particular area (Odum 1971: 5, 8). Figure 5.1 is a simplified model of the ecosystem, illustrating interactions as flows between living organisms and the non-living natural environment. Humans and the human-made environment are introduced as a separate component to show material flows between the socio-economic and ecological systems (flows crossing the boundaries of the socio-economic system). Those flows that have direct influence on human welfare are shown as arrows that hit the humans circle inside the socio-economic system (marked with a plus sign (+) for welfare increases and with a minus sign (−) for welfare impairment). For reasons of simplicity, a number of important internal processes are hidden in 'black boxes' in the flow chart, or neglected: these include, among others, interactions of species within communities, the role of organisms and climate in soil development and human activities in their habitat, i.e., human

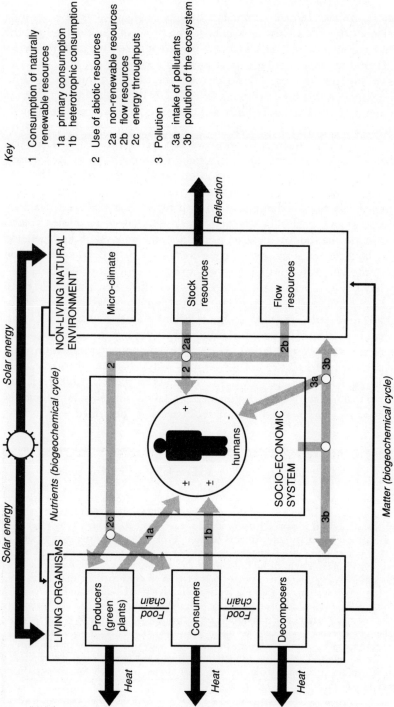

Key

1 Consumption of naturally
  renewable resources

  1a primary consumption
  1b heterotrophic consumption

2 Use of abiotic resources

  2a non-renewable resources
  2b flow resources
  2c energy throughputs

3 Pollution

  3a intake of pollutants
  3b pollution of the ecosystem

Figure 5.1 Man and ecosystem
Source: Bartelmus (1986: 40).

settlements. In fact, human settlements have been regarded as an ecosystem of their own (see the following section). The flows of the chart reflect the fundamental functions of the human environment of resource supply (flows 1 and 2), and waste and pollutants absorption (flow 3).

The throughput of energy flows through the economic system and into bioproductive (agricultural) systems (flow 2c) is of special importance for rural communities. The figure illustrates how the economic system moulds the raw materials into energy forms that can be readily used in agriculture, forestry or fishery. These energy flows can be considered as auxiliary, subsidizing man's transformation of natural growth processes by such means as mechanization, irrigation, fertilization or chemical pest control. The ecological impacts of these activities include the reduction of ecosystem diversity by artificial selection or domestication of plants and animals. The stability of the ecosystem may thus be decreased, making it more vulnerable to pests, diseases and further human impacts. In addition, the removal of the vegetation cover by land clearance and overgrazing creates soil degradation and erosion which in the long run impair the productivity of agriculture and forestry.

Figure 5.1 gives a fairly static picture of man's interaction with ecosystems. To understand the forces that resist or yield to human impacts in ecosystems one has to have some knowledge of ecosystem dynamics and, in particular, of the concept of ecosystem equilibrium. 'Ecosystems are capable of self-maintenance and self-regulation as are their component populations and organisms' (Odum 1971: 33). They tend to resist change and to remain in a state of equilibrium. There are however limits to the adaptability of ecosystems to major disturbances, i.e., their ecological sustainability. Some features of the dynamics and maintenance of ecosystems are described in Box 5.4.

Box 5.4   Evolution, stability and exploitation of ecosystems

The concept of *homoeostasis* refers to the state of an ecosystem operating within limits. An important feature of ecosystem dynamics is that 'really good homeostatic control for the preservation of the ecosystem comes only after a period of evolutionary adjustment' (Odum 1971: 35). Ecosystems typically display a strategy of development, termed *ecological succession*, which culminates in a climax system of high diversity, large biomass, low production and high stability. By contrast, human exploitation is interested in maintaining bioproductive systems in a young high-productivity state. This can be seen as a strategy of maximum production, as opposed to – and indeed conflicting with – nature's strategy of maximum protection or adaptation. It is the main challenge of ecodevelopment to show how nature's strategy of ecological succession can be reconciled with human strategies of maximum production for development. The use of environmentally sound technologies ('ecotechniques') plays a crucial role in this harmonius approach to ecosystem development (see p. 120)

Ecodevelopment features

Most impacts on ecosystems stem from activities of socio-economic development. Ecological criteria were, therefore, introduced above into national development planning and policies. At the local level, ecological criteria have always been applied by traditional agrarian communities which are closely related to the productive rhythm of nature. Examples of full identification of human communities with their physical environment and of their successful management of ecosystems are the hand-built rice paddy terraces on steep mountain slopes in South-East Asia, swidden cultivation in tropical rain forests and traditional herd management by nomads in semi-arid zones.

Aware of these ecological successes and of the significance of ecological factors in the development process, a new local-level planning concept of ecodevelopment was advanced by UNEP. UNEP (1975: para. 100) also offered one of the first explicit definitions of ecodevelopment, as:

> development at regional and local levels ... consistent with the potentials of the area involved, with attention given to the adequate and rational use of the natural resources, and to applications of technological styles ... and organizational forms that respect the natural ecosystems and local sociocultural patterns.

This definition suggests a new regional approach to sustainable development planning in referring explicitly to both ecological and sociocultural habitats. Ideally, an ecoregion would be a homogeneous area of one or more ecosystems that interact with relatively self-contained human activities. Self-contained means that all major economic, social and cultural activities originate in and affect the respective ecoregion, while impacts from and to outside are minimal. Established objectives of a political, administrative, tribal or cultural nature might compel planners to deviate from theoretically desirable ecoregion boundaries. For example, the only politically feasible solution might be to set up a compromise region from the smallest administrative units that cuts across ethnic boundaries. As a consequence, a certain degree of economic, cultural and ecological heterogeneity would be introduced into the region.

An even greater heterogeneity has to be faced if ecodevelopment would be applied to urban areas as proposed by UNESCO (1981). Such areas are usually not self-contained at all. They depend heavily (with the help of a large-scale, capital-intensive infrastructure) on resource flows from the hinterland and far-off regions within and even outside national boundaries. It is probably for this reason and the 'alienation' of urban dwellers from their natural environment that ecological approaches to human settlements planning have not been further pursued in practice. Rather, the common approach to human settlements planning seems to be – after paying dutiful allegiance to 'sustainable development' – the elaboration of key sectorial strategies of land-use, water supply and sanitation, waste management, energy, transportation and construction (UNCHS 1990).

The concept of ecodevelopment planning was picked up on behalf of UNEP in particular by a non-governmental organization, the Centre International de Recherche sur l'Environnement et le Développement. This organization developed numerous pilot studies and established the following main features of ecodevelopment (Sachs 1976, 1980):

1 resource development for the satisfaction of basic needs;
2 development of a satisfactory social ecosystem;
3 rational (non-degrading and non-wasteful) use of natural resources in solidarity with future generations;
4 use of alternative, environmentally sound production procedures;
5 use of alternative energy sources, in particular of the regional capacity for photosynthesis;
6 development and use of ecotechniques;
7 establishment of a horizontal authority, ensuring participation of the population concerned and preventing any plundering of the results of ecodevelopment;
8 preparatory education to create social awareness of ecological values in development.

At the heart of ecodevelopment are items 4, 5 and 6, i.e., environmentally sound technologies. In the local (ecological) context, those technologies have been referred to as ecotechniques. They include a wide range of diverse managerial and technological approaches to ecosystem management, such as biological pest control, aquaculture, non-conventional energy sources (biogas, sun and wind), ecodwelling and traditional medicine (Bartelmus 1986: 54 et seq.). The value of traditional local knowledge and experience in managing natural resources has now found wider recognition. For instance, even the large-scale-projects oriented World Bank refers to the important contributions of local communities in the use of marginal lands, soil conservation, afforestation, irrigation and flood control (World Bank 1992: 93–5).

As at the national and international levels, social aspects of power and income distribution are usually also neglected in the analysis of local (community) and micro-level (household) social organization. As at the national and international levels, the principal reasons are encrusted power structures that strive to conserve traditional privileges based on race, caste or gender. Exhortations of the need to develop a 'satisfactory social ecosystem' (one of the above features of ecodevelopment) or of the vital role of social groups, notably women, in environmental management and development (see for example, principle 20 of UNCED's Rio Declaration) are not enough. Scarce human resources need to be recognized and their services rewarded, even if conventional accounts and economics still consider them as free gifts. A case in point is of course women's involvement in natural resource management (water, fuelwood) or household services (food preparation, education, etc.). Those activities do not obtain an economic value in national income accounts and are thus neglected just as the 'free' natural asset services (see section 2.2, above). Moreover, local élites can be as discriminatory in the absorption of wealth and income as national élites (World Bank 1992: 96).

Improved accounting is of course only a first step. The next step is to break up dominance and discrimination through involvement, participation and empowerment of local groups and individuals. For example, land reform to guarantee gender equity in the distribution of land resources, full valuation and remuneration of 'female capital' services, information and education to change attitudes from dominance to co-operation between sexes have been suggested to 'close the gender gap in development' (Jacobson 1993). Decentralization of central planning and policy-making, and power delegation to local communities are proposed below to this end.

Modelling ecodevelopment

Planning models deal with the assessment of problem areas, the identification of functional relationships between relevant variables and parameters and the description, evaluation and selection of problem-solving strategies. A large variety of models, case-studies and approaches have been suggested according to the particular conditions of ecoregions or other (administrative) areas. The following illustrates only a few approaches that might provide a useful starting-point in developing projects and programmes for particular – usually highly vulnerable – areas.

The input–output approach to displaying and analysing environmental impacts of production activities was developed by Leontief (1970) and applied as an interregional framework of economic–ecological interactions by Isard *et al.* (1972). The identification and description of human and ecological interactions is only a first step in the planning and management process. Input–output matrices can be developed into more sophisticated models of decision-making by assuming functional, linear or non-linear, relationships between inputs and outputs, allowing for constraints such as environmental standards and by introducing an objective function for the selection of an optimal strategy. As described in sections 3.2 and 3.3, models of activity analysis are particularly well suited for selecting optimal programme levels that meet economic and environmental constraints. Their application to ecoregions is discussed by Bartelmus (1986: 51).

Activity analysis is normally static in nature, restricting prediction to externally determined changes of certain control variables and assuming constancy of – usually linear – relationships of variables. By contrast, the system-analytic approach views environmental problems in the broader context of the decision-making process. Dynamic features are introduced explicitly into the decision-making network by means of simulation models. The broad system-analytic context includes the perception of needs, problem definition, problem analysis and modelling, simulation to test alternative strategies, evaluation and selection of alternatives by decision-makers and the implementation and monitoring of operations (Frenkiel and Goodall 1978: 13). Contrary to models of activity analysis, simulation models usually omit the optimization function. It has been pointed out, however, that a 'good insight into the working of the environmental system' will usually provide sufficient indication of what actions should be taken (*ibid.*: 34). System dynamics is usually represented by large sets of difference equations.

A successful application of simulation modelling in a small ecoregion consisting of fragile ecosystems was carried out for the village of Obergurgl in the Tyrolean Alps in Austria (Walters 1978). The model produced predictions of alternative paths of long-term development, which were fully taken into consideration in the village's co-operative development efforts. Figure 5.2 shows the interactions between the four components of the model:

1 recreational demand;
2 population and economic development;
3 farming and ecological change; and
4 land-use and development control.

A similar model was designed for small island states in the Eastern Caribbean (OECS 1989). The model simulates alternative governmental interventions of taxation, zoning, expenditures and migration control. The effects of those measures are tracked in sub-models of investment, land-use, environmental quality, economic activity, demography, government accounts, natural resource stocks and infrastructure.

As discussed above in section 3.1, the sustainability of human populations in a particular area can be formulated in analogy to the concept of the carrying capacity of ecosystems. In principle, activity analysis and simulation modelling have the capacity to incorporate 'exogenously' determined limits in the carrying capacity of local natural systems as further constraints in the sustainability (feasibility) of local-level development. Such analysis requires the specification of targets of human needs or standards of living for the local population and of standards or thresholds for natural resource capacities and environmental impacts in the area (ecoregion).

In contrast to national approaches to sustainable development, such targets and standards can be formulated directly at the local level in a transparent, participatory and democratic manner by the concerned populations. This aspect of immediate, participatory and democratic evaluation of environmental, social, cultural and economic goals, programmes and projects is probably the greatest advantage of a local implementation of multidimensional development.

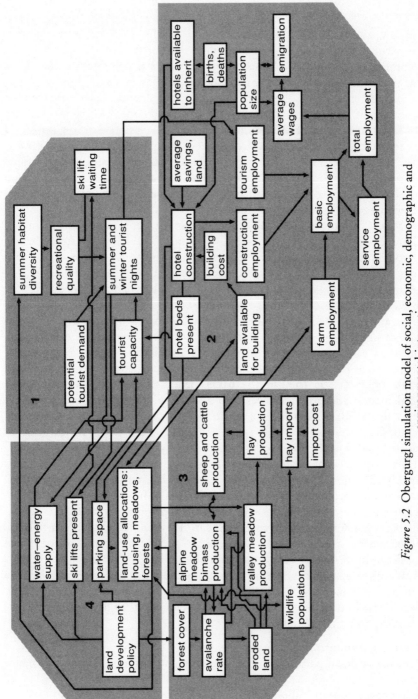

Figure 5.2 Obergurgl simulation model of social, economic, demographic and environmental interactions

Source: Walters (1978).

Decentralization: top-down vs. bottom-up approaches

Empowerment has been considered by UNCED as the key to facilitating the participation of local populations in the sustainable development of their communities. In turn, 'disempowerment' of local communities has been flagged as the root cause of environmental decline in developing countries (Vivian 1991: 18). This reflects the view that the benefits of centrally planned programmes and projects usually do not reach the local poor but are absorbed by the wealthy and powerful in the country or region. As a consequence, a policy of 'selective spatial closure' has been proposed for less developed areas (Stöhr 1981: 45, 46). Such a policy would retain regional control of commodity and factor transfers to and from the region. As a consequence, some amount of central governmental authority would have to be delegated to local organizations.

On the other hand, local planning cannot replace central planning. Complex large-scale, capital-intensive projects, which may span various regions and ecosystems, require central design and control. However, to the extent that such activities affect local conditions, the population and organizations in the region should also become involved in planning and implementing those activities. There are limits, however, to the involvement of local people. NIMBY ('Not-In-My-Backyard') activism, resulting from lack of knowledge about or 'free-rider' attitude towards complex projects and their costs and benefits, may lead to prohibitively (prolonged) costly negotiations between local groups and national project sponsors (World Bank 1992: 15, 95–6). On the other hand, strong correlation between participation and project success can be generally observed (*ibid.*: 95).

Central guidance and control is also needed for the application of national standards to the region. These standards may relate to the distribution of income and wealth, and to the overall quality of the environment. Local efforts, aiming at participation and self-reliance, need therefore to be linked openly to regional and national processes of decision-making. This should be achieved through effective channels of communication and support, and within a coherent framework of integrated planning as discussed above. Such a framework would have to ensure the compatibility of sustainable development strategies not only vertically between central, regional and local administrations but also horizontally among the different ecoregions.

Recently, spatial development planning has been advanced as such an integrative framework (Newels 1993). Such 'regional planning' has typically been carried out as a top-down approach. The idea is to decentralize central planning and management to achieve

1 greater administrative efficiency;
2 increased local participation in development planning and management; and
3 improved resource allocation in line with local preferences.

Some experience with 'integrated regional development planning' has been gained in case-studies carried out under the auspices of the Organization of American States (OAS 1984). The main features of the case-studies are described in Box 5.5. The studies were selected with a view to incorporating environmental considerations into regional development planning for different ecosystems.

It remains to be seen to what extent top-down regional planning can provide an appropriate framework for the active participation of local grass-roots organizations in the design and implementation of projects and programmes. Much will depend on the willingness of central agencies to learn from and support local experience and thus to reconcile 'bottom-up' strategies with their own 'top-down' decision-making. Initial pilot studies of ecodevelopment, undertaken by UNEP and other organizations, were not very successful, mainly because of political factors (Bartelmus 1986: 79, 81). An analysis of case-studies of development from below found that 'the major limiting factor . . . seems to be fear by central (national or international) decision-making centers that they might lose control' (Stöhr 1981: 69).

Box 5.5 Integrated regional development in OAS member states

Dominican Republic

Objective: methodology for development-oriented natural resource inventory;
Features: lack of natural resource data, lack of suitable investment projects;
Results: agricultural zoning methodology;
Pitfalls: excessive data collection and loss of momentum in implementation.

Pilcomayo (Argentina, Bolivia, Paraguay)

Objective: planning the development of a multinational river basin;
Features: multinational study;
Results: identification of 'programme areas' of high development potential in a large region, establishment of regional accounts, use of satellite imagery for mapping;
Pitfalls: creation of multinational institutions before all options were understood by the countries themselves.

San Lorenzo (Mexico)

Objective: solving land-use conflict in urban growth areas;
Features: use of university as counterpart agency, mobilization of local interest and media;
Results: preparation of a land management system;
Pitfalls: use of a university in a practical development study.

Source: OAS (1984: xiii).

The question is how to overcome the resistance of the powerful in a country to giving up control in the process of implementing decentralized ecodevelopment. At least in one major region of the world, governments now seem committed to being more supportive of the establishment of 'autonomous grassroots organizations' (African Charter for Popular Participation in Development, Arusha 1990). As indicated above, similar commitments to community-driven approaches have been made in UNCED's Agenda 21 (United Nations 1993c: chs 26, 28).

5.3 Implementation: towards a national action plan for sustainable development

Environmental dimensions of central plans and programmes

In most developing countries, the traditional and popular approach to socio-economic development is the establishment of a national medium-term development plan. The plan commonly specifies the major societal objectives and indicates how they can be achieved at various planning levels. Central planning agencies typically prepare the plan in a top-down approach which distinguishes comprehensive or aggregate plans for the whole nation from sectoral and regional programmes and project development. Project development is supposed to translate the overall plan into concrete developmental action.

In most – capitalistic – industrialized countries, national plans have been typically indicative only, avoiding central prescription and enforcement. In Japan, for instance, eleven economic plans have been formulated to

1 provide the perspective of the desirable and feasible orientation of the economic society;
2 identify targets and means of priority policies; and
3 to set guidelines for home economies and corporate activities (Government of Japan 1991).

The integration of environment and development was put on the international agenda by the United Nations Conference on the Human Environment in 1972. Since then, the typical approach at the national level has been to *add* an environmental chapter in national development plans. However, the short-sighted nature of medium-term plans, which only look about five years ahead, has been the principal obstacle to the effective and permanent integration of the environment into the national socio-economic concerns and objectives. Although some symptoms of environmental degradation and depletion do appear during such a time span as a result of past and current economic activity, many environmental and related health effects of present activities will only be felt by future generations. Short-lived administrations cannot be expected to be overly concerned with situations that lie beyond the next few legislative periods. Such an attitude has led to what has been aptly termed 'colonizing the future' (Kothari 1980: 435–6).

It is to the credit of the WCED that it has brought about a change in this attitude towards long-term planning by pointing out the manifold inter-dependences of current socio-economic activities and long-term environmental effects. This change in attitude is also reflected in national reports submitted to UNCED. However, even a superficial review of those reports reveals that actual or proclaimed efforts to implement sustainable development usually confine themselves to extending the time horizon for environmental plans but fall short of their integration in a comprehensive sustainable development plan for the nation. They do elaborate, on the other hand, a number of important instruments for implementing environmental objectives through, in particular:

1 legislation and regulation;
2 project and programme appraisal;
3 science and technology;
4 institution building;
5 supportive measures of environmental data development and awareness building; and
6 financing/budgeting.

Those instruments are important means of plan implementation that could and should be built into comprehensive action plans for sustainable development. They are therefore briefly reviewed in the following, before discussing two examples of how to establish such an action plan in a developing and an industrialized country.

Means of plan implementation

Legislation and regulation Legislation and regulation provide the immediate means of translating the above-described principles, criteria and standards of sustainable economic growth and development into binding action by individuals and social groups. This can be achieved either ad hoc or, preferably, based on a consistent policy framework or plan. In the past, most legislative environmental action was rather spontaneous. It responded to particular environmental phenomena as they reached levels that could neither be hidden from nor accepted by society.

Laws and regulations need to be established not only for command-and-control action but also for the application of market instruments, based on legally stipulated standards in order to ensure compliance and enforcement. This applies not only to national environmental standards and economic targets but also to international conventions and obligations, ratified by national parliaments thus becoming national law. The legislative process also provides the necessary visibility to monitor the implementation of national goals of sustainable development and international commitments as, for instance, developed at UNCED (see United Nations 1993c: ch. 6). To date, many environmental laws have been passed. However, their implementation has been frequently frustrated by lack of 'effective instrumentation and institutional support' (UNDP 1992b: 12). Some aspects of such support are discussed below.

Project design and appraisal As pointed out above, medium-term plans typically consist of sectoral and regional programmes that are translated into concrete action by development projects. Cost-benefit, cost-efficiency or risk analyses are widely used to assess and compare the net benefits of projects for their design and selection. Environmental costs of public projects are usually addressed in Environmental Impact Assessments (EIAs) and statements. Box 5.6 provides a definition of EIA and a list of its main components and techniques.

As described above, private decision-makers can be prompted to incorporate environmental costs into their budgets. Public decision-makers,

---

Box 5.6   Environmental Impact Assessment (EIA)

1   *Definition*: EIA is an analytical process that systematically examines the possible consequences of the implementation of projects, policies and programmes. Its main objective is to provide decision-makers with an account of the implications of alternative courses of action before a decision is made and to then alter final project design (Clark, Bisset and Walthern 1980).

2   *Activities*: most EIAs involve the

- identification of impact;
- measurement of impact;
- interpretation of the significance of the impact;
- display of the results of assessment;
- development of ameliorative measures to eliminate or minimize adverse impacts;
- identification of appropriate monitoring schemes.

3   *Techniques*: typically the following techniques have been applied for EIA:

- mapping and overlay charts;
- checklists of potential environmental impacts;
- matrices of project actions and environmental impacts;
- flow charts, based on systems analysis;
- other models of economic–environmental interaction.

Source: UNDP (1992a: 57–8).

---

on the other hand, have to follow legislative or administrative rules of project/programme appraisal which makes them the prime target of comprehensive (allowing for social costs) cost–benefit analyses (see Box 2.6 above) of projects and programmes. Such cost incorporation can be achieved in principle by manipulating the discount rates of project returns or benefits for purposes of maintaining environmental capital. Considering the above conclusions on the efficacy of discount rate manipulation for environmental purposes (see p. 97), full environmental costing of project packages that include 'shadow projects' to compensate for the environmental damage of other projects has been suggested as an alternative (Pearce, Markandya and Barbier 1990: 27–128).

However, there does not seem to be a system or mechanism ensuring that social environmental values are translated consistently, i.e., without contradictions or conflicts, into national and regional objectives, sectoral targets and from there into criteria of project design and evaluation. As a consequence, the above tools of project appraisal should be considered as piecemeal at best. Moreover, the ability and willingness of central planners to take into account environmental criteria for projects that refer to marginal lands and marginalized populations can be questioned. It can also be assumed that environmental concerns will not be given high priority if they enjoy international financial support, thus already approved 'in principle'. Considering also that the tools of EIA are only applied to selected projects and programmes, they cannot be considered as a systematic approach to the implementation of sustainable development. Nonetheless, attitudes of both recipients and donors of technical assistance appear to have changed and integrated planning and action plans for sustainable development are more widely accepted. This might provide now for a more serious and expanded application of project appraisals in line with overall social goals, targets and standards.

Science and technology   Ambiguities in the scientific evidence for the non-sustainability of human activities were illustrated in the first chapter (section 1.3). They pointed to a lack of precision in the assessment of the state of sustainable development. One of the regional preparatory conferences for UNCED also admitted the lack of scientific understanding of natural systems and formulated what has become known as the precautionary principle in responding to scientific uncertainties. The principle calls for environmental action even when not all facts can be fully assessed, since 'it is better to be roughly right in due time than to be precisely right too late' (Ministerial Declaration on Sustainable Development in the ECE region adopted at the Bergen Conference on 'Action for A Common Future', Bergen, 8–16 May 1990).

This principle was also recognized at UNCED. The conference called not only for developing a better scientific understanding of complex ecological systems but also of their equally complex interaction with human systems, i.e., their carrying capacity and resilience. For rational policies and strategies of sustainable development, improved interaction between the sciences and policy-making needs to be achieved through better communication among scientists, decision-makers and the general public (United Nations 1993c: ch. 35). Such interaction is a prerequisite for assessing the feasibility of sustainable development which was defined above in terms of the interface between social and environmental norms and programmes of socio-economic development.

The development of readily applicable environmentally sound technologies should be the result of such interaction between political guidance and scientific research and development:

> Environmentally sound technologies protect the environment, are less polluting, use all resources in a more sustainable manner, recycle more of their wastes and products, and handle residual wastes in a more acceptable manner than the technologies for which they were substitutes.
>
> *(ibid.*: para. 34.1)

This definition by the UNCED indicates the importance of technological solutions to overcoming non-sustainability of current production and consumption processes. As described above, such technologies widen the feasibility space for socio-economic activities and increase their levels and options.

Incentives for discarding environmentally adverse and applying environmentally sound production and consumption patterns and techniques were described above (section 4.3). In the case of developing countries, it is widely held that such incentives are not enough but need to be supported by favourable access to and transfer of environmentally sound technologies. On the other hand, the above discussion of indigenous 'ecotechniques' also indicated significant contributions of traditional knowledge and techniques to the sustainable management of local ecosystems that could be combined with 'modern' technologies for industrialization.

Recent reports of the United Nations advocate the promotion of technology co-operation as a means of overcoming the 'barriers' to technology transfers (United Nations 1992b: 5; see also Box 5.7). Technology co-operation is considered to be 'a qualitatively higher level of technology transfer' which

1 reflects a broader system approach to technology development and use;
2 promotes co-operative research and technological development in developing countries;
3 interfaces the transfer of knowledge and know-how with domestic innovation; and
4 focuses transfers on technologies that can be rapidly diffused in developing countries.

Box 5.7 Barriers to technology transfer

The following are key items from a list of 'barriers to the transfer of environmentally sound technologies to developing countries at favorable terms' (UNCTC 1991: annex II).

Barriers cited by *transnational corporations*:

- inadequate profits due to the small scale of markets, lack of information, and high costs of training recipients;
- import restrictions;
- unfair competition in the absence of national and international (environmental) standards and regulations;
- inadequate protection of patents.

Barriers from the point of view of *developing countries*:

- high prices of obtaining and high costs of implementing new technologies;
- price advantages of 'dirty' technologies;
- lack of information about options;
- bias against indigenous technologies;
- incompatibility of international standards and technologies with national and local conditions, needs and priorities.

Institution building The integration of objectives, needs, policies and programmes across different disciplines, institutions and social groups, acting at different regional levels, calls for new institutional arrangements that facilitate co-operation and participation. According to UNCED (United Nations 1993c: ch. 23), real social partnership between central government and major groups of women, children and youth, indigenous communities, non-governmental organizations, local authorities, workers and business, the scientific/technological community and farmers needs to be developed in support of common efforts for sustainable development.

The roles and capabilities of some of those groups, notably local communities, industry and consumers, were discussed above (section 3.1 and 5.2), as was the significance of the institutional set-up in sustained growth and development. However, the different social, cultural and political systems, conditions and priorities, and lack of experience with actually establishing participatory institutions for sustainable development do not permit generalizing about the most appropriate approach to facilitating social partnership. Instead, the case-studies, discussed below, of the implementation of a 'green plan' and a 'national sustainable development strategy' will describe how two countries set about actively involving various groups and institutions.

Supportive measures: data development and awareness building Sustainable development planning and policies address, besides economic (growth)

131

variables, social, demographic and environmental ones. Integrated modelling of economic activities and other non-economic programmes and processes was proposed for the broad multidisciplinary analysis of sustainable development. Such models require large sets of variables and corresponding statistics and indicators for their quantification.

Systems of (environment) statistics and indicators aim at measuring these variables in an integrative, or at least comparable, fashion. They are to provide a synthetic picture of the state and trend of the environment and its links to human socio-economic activities. For example, the Statistical Division of the United Nations is actively promoting the application of methodologies of environment statistics, organized in a Framework for the Development of Environment Statistics (FDES) (United Nations 1984, 1988 and 1991a). The Framework links social, demographic and economic statistics with data on environmental impacts and social responses to those impacts. Databases, organized in this manner, can provide the 'raw' statistics for physical environmental (resource) accounts, underlying integrated environmental–economic accounts. They also go beyond such accounting by producing the non-monetary data required for the assessment of the – non-monetary – aspects of sustainable development (for example, of poverty, health, shelter or cultural values) (Bartelmus 1992b). Definitory descriptions of environmental statistics and indicators are given in Box 5.8.

Box 5.8 Environmental statistics and indicators

Based on the FDES (United Nations 1984) *environment statistics* can be defined as 'statistics that describe the state and trends of the environment, covering the media of the natural environment (air/climate, water, land/soil), the biota found within these media, and human settlements'. Environment statistics are integrative in nature, measuring human activities and natural events that affect the environment, the impacts of these activities and events, social responses to environmental impacts, and the quality and availability of natural assets.

The construction of *environmental indicators* consists of a logical selection and/or combination of available environmental data. The problems of selecting and aggregating 'social indicators', as described in Box 3.1 above, apply also to the construction of environmental indicators or compound indices. According to the OECD (1991), indicators are series selected from a larger database with a synthetic meaning and specific purpose. Therefore, there is no universal set of environmental indicators; rather, there are sets of indicators responding to specific conceptual frameworks and purposes. For example, there are three types of indicator sets, currently being developed by OECD, to

1 measure environmental performance (level and change of environmental quality);
2 integrate environmental concerns in sectoral policies; and
3 integrate environmental concerns in macroeconomic policies.

The need for reducing the information overload of large sets of statistics and monitoring data led to the search for indicators of sustainable development by UNCED (United Nations 1993c: para. 40.6). Such indicators would include not only the above-discussed monetary measures of Environmentally-adjusted National Income and Domestic Product and their distribution, but also physical indicators of the efficiency of natural resource use and the effects of environmental impacts on human health and well-being.

Data produced by integrated accounting and statistics need to be translated into public information. On one hand, improved information about environmental effects might bring about changes in lifestyles such as 'green consumerism' (Pearce, Markandya and Barbier 1990: 155). On the other hand, strategies of increasing awareness, education and participation are a prerequisite for the acceptance of policies aiming at the incorporation of environmental costs into the accounts and budgets and, more generally, into the decision-making process of enterprises and households. Environmental education, training and awareness campaigns might also be the only means to obtain the participation of communities that are outside the reach of conventional top-down policies. This situation arises when those communities are either neglected by central government or are independent enough to resist governmental control and intervention (see, for instance, the case-study of PNG, below).

Finance and budgeting Implementing a new integrative approach to sustainable development will require additional efforts in terms of both human and financial resources. One frequently heard argument is that in the case of environmental non-sustainability of economic growth and development, the cost of inaction outweighs the financial costs of implementing sustainable development. Integrated accounting has been designed to assess the costs of action and inaction and can thus provide the basis for projections of those costs. In principle, additional resources will have to be found through reallocation from lower-priority development programmes to higher-priority ones. This might be the ultimate proof of a country's commitment to sustainable development in terms of its willingness to pay for such development. To some extent, additional funds are built-in results of certain instruments such as fiscal measures (elimination of environmentally detrimental subsidies or introduction of effluent charges) discussed above (section 4.3 and Box 4.8). Other instruments of fund-raising for sustainable development are voluntary contributions resulting from changed awareness and attitudes regarding environmental questions, or 'peace dividends' from disarmament.

There was consensus, however, at the UNCED that the implementation of its huge blueprint for sustainable development, the Agenda 21, will require the provision to developing countries of substantial new and additional financial resources (United Nations 1993c: para. 33.10). Bilateral or multilateral technical assistance from other countries or international organizations would therefore have to provide support to those countries which, due to their difficult economic situation, are not able to ensure the sustainability of their development on their own. Technical assistance should include

1 the transfer of environmentally sound technologies and know-how;
2 debt relief (possibly linked to environmental action, see Box 4.7); and
3 direct financial assistance for the implementation of projects that aim either at reorientation of unsustainable policies or at the clean-up of past environmental degradation.

International arrangements and commitments in this regard are discussed in the concluding chapter.

An interesting mechanism for ensuring a continuous flow of financial support for long-term sustainable development has been proposed recently, *viz.*, the establishment of a trust fund (Corbin 1993). Such a fund would free the recurrent budget of developing countries from financing the maintenance of its capital stock. Currently, allocations to capital maintenance are generally insufficient, especially when capital returns are low or long-term only. The result is rapid decline of the capital stock. The notion of a trust fund could be extended to the financing of the maintenance of natural capital. An environmental trust fund would be particularly appropriate since no allowance for natural capital maintenance is usually made in conventional accounts. Even the new integrated accounting methodologies can only alert to the necessity of such a fund since they account for 'hypothetical' (imputed) costs only (see section 2.3, above).

Case-studies: Canada's Green Plan and PNG's Sustainable Development Strategy

The idea of integrating environment and development in policy-making and planning is not new. What may be new, is the political will to implement this integration. The consensus reached at UNCED seems to reflect such a change in attitude in both governments and non-governmental organizations. However, even the best intentions will have to face the realities of differing priorities and values in society. Conflicts may thus emerge when concrete action is taken to implement sustainability criteria by different groups and different governmental authorities. About a year after the Rio Conference, international organizations, governments and non-governmental organizations, as well as individuals, were still in the process of taking stock of the results of UNCED, examining the implications they may have for their policies, plans, strategies and behaviour.

One such attempt by a developing country (Papua New Guinea) at stock-taking and drawing up conclusions in terms of a national action plan, in which the author was personally involved, is described below. An earlier approach to sustainable development by an industrialized country (Canada) is also outlined. The country had actively participated in the path-breaking work of the WCED on sustainable development, well before the Earth Summit agreements were reached. The two examples are to illustrate alternatives for setting out to make sustainable development a national reality.

Canada's Green Plan   In its report to UNCED, the Canadian government claimed that its Green Plan 'puts Canada firmly on the path toward sustainable development' and 'provides a long-term strategy for making the fundamental changes that will be needed to bring environmental considerations into the mainstream of the day-to-day decision-making of Canadians' (Government of Canada 1991: v). Crucial in bringing environmental criteria into the decision-making process have been public consultations in which over 10,000 Canadians presented their views to the government. Those consultations are seen as a prerequisite to forging partnerships among different sectors of society that may have different and competing interests in the economy and the environment. It may well be that this – continuing –

Box 5.9   Green Plan, Canada – goals and selected initiatives

1   *Life's Three Essentials: Clean Air, Water and Land* (Can. $850 million)
   – the health and environment action plan;
   – smog emissions control;
   – national waste reduction plan.
2   *Sustaining Our Renewable Resources* (Can. $350 million)
   – sustainability in agriculture, fishery and forestry.
3   *Our Special Spaces and Species* (Can. $175 million)
   – completing the national parks systems;
   – the national wildlife strategy.
4   *Canada's Unique Stewardship: The Arctic* (Can. $100 million)
   – the Arctic Environmental Strategy.
5   *Global Environmental Security* (Can. $575 million)
   – the national action on global warming;
   – expanding acid rain controls.
6   *Environmentally Responsible Decision-Making* (Can. $500 million)
   – partnership programmes;
   – environmental information;
   – harnessing market forces.
7   *Starting in Our Own House* (Can. $275 million)
   – environmental assessment;
   – environmental stewardship.
8   *Emergency Preparedness* (Can. $175 million)

Source: Government of Canada (1990, 1991).

---

consultative process and its resulting partnerships and co-operative efforts might prove to be more important than the documentation it produced, i.e., the Green Plan itself.

The institutional mechanisms to facilitate consultations were developed on the basis of recommendations of a National Task Force on Environment and Economy. The Task Force, which brought together for the first time environment ministers, business executives, environmentalists and academics, recommended the establishment of Round Tables on the Environment and Economy by the prime minister and the premiers of the provinces. The National Round Table consists of leaders from all sectors of

---

Box 5.10 National Action Plan for Sustainable Development, Papua New Guinea – major themes and priority actions

1 *Revitalizing growth with sustainability*
   Acceleration of sustainable development;
   Integration of environment and development in decision making.
2 *Sustainable living*
   Combating poverty;
   Changing consumption patterns;
   Population;
   Health.
3 *Human settlements*
   Human settlements;
   Urban water supplies;
   Solid waste management;
   Urban pollution and health.
4 *Efficient resource use*
   Land;
   Freshwater;
   Energy;
   Agriculture and rural development;
   Mountain development;
   Development of coastal areas;
   Development of islands;
   Biological diversity;
   Biotechnology;
   Managing fragile ecosystems.
5 *Global and regional resources*
   Atmosphere;
   Oceans and seas;
   Sustainable living marine resources.
6 *Managing chemicals and waste*
   Toxic chemicals;
   Hazardous waste.
7 *People participation and responsibility*
   Education, public awareness, training;
   Strengthening the role of social groups.

---

society and provides a forum for continued dialogue, formation of new partnerships and evaluation of policies. Similar round tables were established in ten provinces and territories.

The Green Plan (Government of Canada 1990) itself reflects a governmental commitment, supported by over 40 federal departments and agencies, to solving environmental problems and implementing economic development that is sustainable. The Plan sets out national environmental goals and targets, as well as over 100 'initiatives' (programmes/projects) to achieve the goals and targets. Some of those initiatives are shown in Box 5.9. The box also indicates the additional (over and above existing federal environmental expenditures) funds allocated to each goal of the Green Plan.

Papua New Guinea: towards a sustainable development strategy Papua New Guinea (PNG) has been one of (if not *the*) first country to initiate a process of translating the voluminous recommendations of UNCED into a national action plan. This may be due to the fact that PNG's constitution already enshrines environmental concerns and the sustainability of its natural wealth in its fourth goal. The goal refers to 'Papua New Guinea's natural resources and environment to be conceived and used for the collective benefit for us all, and to be replenished for the benefit of future generations.'

The government, through the Departments of Environment and Conservation and Foreign Affairs thus organized a Post-UNCED Seminar in November 1992 with the objectives of

1 defining the concept of sustainable development as applicable to PNG;
2 recommending the implementation of agreed principles and policies for different sectors;
3 developing a framework for a National Action Plan; and
4 recommending to the government priority areas for follow-up action.

The seminar aimed at the involvement of governmental and non-governmental organizations and groups in the country. It revealed, however, the difficulties of obtaining a broad-based consensus on a complex multi-disciplinary matter within a one-week period. The discussion of a draft Framework for a National Action Plan for Sustainable Development, whose seven 'themes' are shown in Box 5.10, focused therefore only on four 'elements' of the framework. Those elements are:

1 the social and economic dimension;
2 conservation and management of resources and development;
3 strengthening the role of major groups; and
4 means of implementation.

Probably more important – at that stage of development – than the substantive discussion are the institutional arrangements for setting in motion the process of developing a National Sustainable Development Strategy (NSDS). It is intended to establish an NSDS Steering Committee which would be serviced by a Technical Task Force. The Task Force should include a natural resource economist, a legal expert, a communications expert and a social scientist. Together, they would carry out the substantive work through a broad-based consultative process.

Perhaps the most striking impression from the seminar is its focus on local-level landowner participation. This appears to reflect the reality of customary land tenure and an undeveloped monetized market system in PNG, together with relatively weak central powers – regarding natural resource management. As a consequence, overall national planning and policies for sustainable development might play a secondary role in implementing the NSDS, while integrated local-level strategies, for example, of rural development, might be the key to sustainable resource management and environmental care. The first two articles of the declaration of a landowners association, presented in Box 5.11, reveal the pride and seriousness applied to environmental concerns by local tribal organizations and traditional landowners.

One lesson for countries with similarly undeveloped monetized markets might be the relatively limited contribution of central government to the implementation of sustainable development. The only means of instilling sustainability criteria into local-level planning and strategies appears to be campaigns of awareness-building through information dissemination,

---

Box 5.11 The Declaration of Environmental Policy Regulation, Namea Landowners Association Incorporated

1. This association . . . solemnly declares now that it will always
   1.1 assure for all Abau people's safe, healthful, productive and aesthetically and culturally pleasing surroundings;
   1.2 attain the widest range of beneficial uses of environment without degradation, risk to health or safety, or other undesirable and unintended consequences;
   1.3 preserve our important unwritten histories, cultures, and natural aspects of our heritage and maintain, wherever possible, an environment which supports diversity and variety of our individual choices;
   1.4 achieve a balance of standards of living and wide sharing of life's amenities; and
   1.5 enhance the quality of renewable resources and approach the maximum attainable recycling of depletable resources.
2. The Namea Landowners Association recognizes that each person should enjoy a healthful environment and each person has a responsibility to contribute to preservation and enhancement of the environment.
   (First two out of 12 articles, declared at Waigani this 22nd Day of October, 1990)

---

extension services and education. In PNG, the key actors in translating international and national guidelines and recommendations into local-level action are the non-governmental organizations, in particular the church and women's groups. Those groups are best equipped to reach local communities in a country of impenetrable forests, swamps and rugged mountains where communication and transportation are major problems.

On the other hand, the national seminar also considered that most of the environmental problems in the country were 'imported', for example, through external demand for the natural resources of minerals, timber and fish. Both depletion of those resources and environmental impacts from their exploitation are usually brought about by the mining, logging and fishing activities of foreign corporations. Governmental monitoring and control of those activities could reinforce local-level environmental conservation and protection efforts and facilitate fair negotiations between traditional landowners and foreign companies. Such negotiations might indeed lead to optimal uses of environmental assets according to the so-called Coase theorem (see, however, for a general critique of the theorem Pearce and Turner 1990: 73–83).

The two examples of making the transition from rhetoric to action indicate not only problems and solutions in obtaining a broad-based consensus on how to implement sustainable development; they also show that the *process* of reaching the consensus generates widespread awareness of environmental questions. In the long run, this process might generate its own dynamics of dealing with environmental issues – within or outside established planning and policy-making structures. Clearly, countries can learn from their respective experiences. International organizations should encourage the exchange of information and expertise in this regard (see also Chapter 6).

6

OUTLOOK: FROM NATIONAL TO GLOBAL COMPACTS

The concluding chapter argues that the response to the assessments of the opening chapter has to be found in global consensus and co-operation. The process of consensus-building has culminated in the Rio Summit (UNCED) whose outcomes are briefly reviewed. The book ends with a plea for stronger supranational institutions for the effective implementation, monitoring and further development of the principles, recommendations and conventions agreed upon in Rio.

6.1 Towards global sustainability

'Traditional forms of national sovereignty are increasingly challenged by ecological and economic inter-dependence' (WCED 1987: 261). Such interdependence is apparent in the cases of deteriorating global commons (atmosphere/climate, outer space, oceans and the Antarctica), use of shared natural resources and transboundary pollution. Cross-boundary effects are caused by a multitude of human activities, creating together global effects that, in turn, may affect national and local environmental systems and socio-economic activities. Examples are local deforestation and point-source emissions that may create global atmospheric changes which in turn affect bioproductive systems at different geographical levels. As indicated in the last column of Table 4.1, international agreement and co-operation are needed to address comprehensively such interaction.

Strategies of global sustainable growth and performance

The maintenance of natural assets in sustainable economic growth applies at the international level to assets which are:

1 shared by neighbouring countries, for example, in the case of cross-boundary freshwater or marine resources;
2 owned by all nations in the case of global commons;

3 substituted by natural resource imports in particular from countries with less stringent environmental standards; or

4 affected by cross-boundary environmental effects, for example, from acid precipitation or waste dumping outside the national territory.

An objective analysis of the actual and potential depletion and degradation of shared and global resources and their effects on national economies requires global or regional data compilation on and modelling of economic–environmental interactions. A promising attempt is currently being made to update and revise a global input–output model. The model aims at providing the physical underpinning for projections of the sustainability of development at the international level. So far, however, the focus has been on energy use and associated emissions only (United Nations 1992e). A broader analysis should help to determine if global natural resource limitations require an international effort to limit global economic growth, or if allocative policies of international resource use and trade (regulation) are necessary or sufficient to achieve global sustainability of economic growth.

Trade liberalization and full-cost pricing through incorporation of all environmental costs were described above as a means of providing an optimal distribution of production among countries according to their comparative advantages in endowment with natural, human, produced and institutional capital. However, the negotiations in the context of the UNCED have revealed a risk of misusing multilateral environmental agreements for the (re)introduction of 'unjustified' restrictions on trade (see section 4.2, above). As a consequence, the international distribution of production might be far from optimal and might be concentrated (relocated) in countries with less stringent environmental standards and costing. Compensation for countries which 'export' sustainability in this manner has been proposed (Pearce, Markandya and Barbier 1990: 47).

Possibilities of applying the above-described economic instruments at the international level need also to be explored in this context. International taxation of natural resources and especially of energy use, based on the user-pays principle, has been proposed. Similarly, the polluter-pays principle would be the justification for international pollution charges or for the international trading of 'pollution rights' (Dasgupta and Mäler 1990: 35). The implementation of international economic instruments, especially with a built-in 'automaticity' in response to environmental impacts, would require global consensus on natural resource capacities and uses, as well as on the levels, pathways and effects of pollution. International conventions and agreements would pave the way for such consensus, at least in selected areas such as fossil fuel use and CO_2 emissions.

Global sustainable development: implementing international conventions

International conventions could play a key role in the analysis of the global sustainability of development. A multidimensional concept of sustainable development was introduced above (section 3.2) to integrate non-economic aspects of development with economic ones by means of a normative framework of standards and targets. At the international level, such a framework would have to address, apart from global resource capacities, international ('harmonized') pollution standards and the issue of international equity in the distribution of income and wealth (including natural wealth). Additional global concerns such as environmental and economic security could also be considered as further targets of or constraints to global development.

International conventions such as the Montreal Protocol on Ozone Depleting Substances, the Basle Convention on the Transboundary Movement of Hazardous Wastes, the Convention on International Trade in Endangered Species or the recently negotiated conventions on climate change and biological diversity (see Box 6.2, below), aim at setting and/or harmonizing environmental targets or standards. A consistent and comprehensive set of international agreements could provide a global normative framework for the implementation of comprehensive integrated development. Once international standards in both the economic and environmental fields have been agreed upon, they could be introduced into global models such as the above-mentioned input–output model for the determination of 'feasible' international development strategies. At the same time, such an international framework would permit the reduction of competitive distortions among countries due to differing national environmental standards and regulations.

However, there does not seem to be any mechanism of regularly monitoring compliance, let alone of enforcement. It is thus an open question to what extent countries actually implement the stipulations of agreed conventions and protocols, even if those agreements are ratified by national parliaments. Non-compliance can thus be suspected especially in countries where high costs of implementation are involved. For instance, the phasing out of CFCs used in refrigeration and air-conditioning would raise the price from $0.30 (for 500 g of CFC) to about $3.0 (for its substitute HFC-134a). Such a cost increase could thwart fledgling industrialization efforts in some developing countries, for example, India and China where those industries are or could become significant (Bhalla and Bifani 1992: 143).

Differences in the endowment with natural resources and waste/residual absorption capacities can be the ground for both trade in those resources and (environmental) services, and for tension among countries. The latter may occur if national shortages in those endowments reflect insecurity on the part of countries that have to import natural resources and to export

142

wastes and residuals. Examples of such tensions are conflicts such as the intervention in Iraq that are, at least partially, motivated by ensuring access to strategic resources. Other examples are transboundary deposits of acids in Europe and North America or ecological refugees in East Africa. International regimes, facilitating environmental or resource security, increased regional self-reliance through diversification and stabilization of resource prices have been suggested to increase global economic and environmental security (Opschoor 1989).

The goal of equity at the international level, i.e., the approximation of standards of living between industrialized and developing countries, might only be achieved by radical changes of production and consumption patterns. The need for correcting imbalances in global patterns of production and consumption has been recognized by the UNCED. The 'excessive demands and unsustainable lifestyles among the richer segments' are denounced in Agenda 21. The answer is 'reorientation of existing production and consumption patterns that have developed in industrial societies and are in turn emulated in much of the world' (United Nations 1993c: ch. 4). It is doubtful, however, whether market incentives and corresponding price signals can achieve this correction alone. Strong international commitment with additional regulatory action is needed. Also, the potential consequences of such corrective action for the sustainability of development in both developing and industrialized countries has to be assessed. The capabilities of individual countries to participate in international efforts, from both a technical and financial point of view, need to be appraised with a view to sharing the environmental and economic costs involved.

Developing countries generally maintain that developed countries are responsible for most of the global environmental problems and thus also for dealing with these problems. Considering further that the alleviation of dire poverty is a *sine qua non* for any type of development, leads to the conclusion that funds for implementing sustainable development need to be found over and above the costs of poverty alleviation. For developing countries the costs of 'socially necessary' (meeting basic development objectives) economic growth have been estimated at an additional net capital inflow of $40 billion in 1990 ($60 billion by the year 2000). If requirements for environmental protection are added, an additional $20 billion in 1990 ($80 billion in 2000) would be needed (Jayawardena 1991: 8). Cost estimates by the UNCED Secretariat of implementing UNCED's Agenda 21 are much higher. International funding of Agenda 21 would require, on average, about $125 billion annually up to the year 2000 (United Nations 1993c: para. 33.18). The total costs of implementation, including national expenses, are estimated at four to five times this amount.

Box 6.1   International funding mechanisms

Global Environment Facility (GEF). The GEF is jointly managed by UNDP, UNEP and the World Bank. It is a three-year experiment that provides grants for investment projects, technical assistance, and – to a lesser extent – research. The Facility has Special Drawing Rights of US$1.3 billion to commit over the three-year period. The purpose is to assist developing countries to deal with four global environmental problems, namely:

1 global warming;
2 pollution of international waters;
3 destruction of biological diversity; and
4 depletion of the stratospheric ozone layer.

Thus, projects that are deemed to benefit the global environment, as distinct from the local environment, only qualify for funding under the GEF (UNDP, UNEP and the World Bank 1991).

Capacity 21. The objectives of this UNDP programme are to:

1 formulate, on the basis of existing plans, sustainable development strategies;
2 formulate a national Agenda 21;
3 identify capacity building requirements for the implementation of the national Agenda 21; and
4 formulate programmes and projects to meet the capacity building requirements.

By mid-1992 about US$18 million was raised. It is hoped to reach $500 million by the end of 1995 (UNDP notes and briefs).

Peace dividend. As a result of falling military expenditures, UNDP (1992b: 85–7) estimates that an immediate windfall of $158 billion for industrialized and of $119 billion for developing countries has already become available. Of course those amounts would not automatically be invested in natural resource conservation and environmental protection, but would have to compete with other social goals and priorities.

6.2 Earth Summit: an agenda for the twenty-first century

However, no final commitments to establishing one or more green funds could be reached at UNCED. In fact, negotiations for 'new and additional' financial resources were the most difficult and protracted. In the end, all that could be presented in the chapter (33) on finances was a listing of actual and potential funding sources and mechanisms (see Box 6.1), notably:

1 the Global Environment Facility (GEF);
2 different multilateral financial institutions; and
3 a few 'innovative' instruments such as the above-described debt swaps (see Box 4.7), international economic instruments and the use of 'peace dividends' from disarmament.

The GEF has drawn considerable criticism regarding the way it had been managed and its focus on global environmental phenomena. Agenda 21, therefore, requested to expand its scope and coverage and to ensure 'transparent and democratic governance' as well as better 'predictability' in the flow of funds from developed countries (United Nations 1993: para 33.14). Aware of those deficiencies, UNDP has recently proposed the establishment of a Capacity 21 programme addressing the areas that are not covered by the GEF. The idea is to build up the capacities of developing countries in implementing Agenda 21 to achieve sustainable development (see Box 6.1).

Many participants and observers at UNCED were and are inclined to measure the success or failure of the Conference in terms of the willingness to pay of industrialized countries. Others have pointed out that UNCED was aiming at a consensus on priority concerns, principles and action to address those concerns through international co-operation. While pledging financial contributions might indeed reflect commitment of governments to solving global problems much, if not most, of problem-solving has to start at home by implementing national and local policies of sustainable development.

Obviously, views on the results of UNCED differ widely. They range from:

1 describing its main output, the Agenda 21, as 'the most comprehensive, the most far-reaching and, if implemented, the most effective programme of international action ever sanctioned by the international community' (closing statement by the Secretary-General of UNCED, M.F. Strong) to
2 considering the Conference as 'a failure of historic proportions' (Greenpeace summary critique of official UNCED results).

It is of course too early to assess the effects on national and international policies of UNCED documents that aim at developing co-operation at national and international levels for the rest of the current and much of the twenty-first century. However, the following brief review of the more tangible results of UNCED gives an impression of what it should be possible to attain on the basis of the agreements reached in Rio. Figure 6.1 provides a synopsis of the results of UNCED. Those results are a declaration, an action plan (Agenda 21), the adoption of two conventions on biodiversity and climate change and a statement of forest principles. A summary of the two conventions and the forest principles is given in Box 6.2.

The Rio Declaration presents 27 principles that are a watered-down version of an originally planned 'Earth Charter'. Frequent use of 'should' instead of the more binding 'shall' and qualifiers such as 'according to their capabilities' (for example, in reference to the precautionary principle) reflect what has been termed a 'lightweight' declaration by NGOs. Nonetheless,

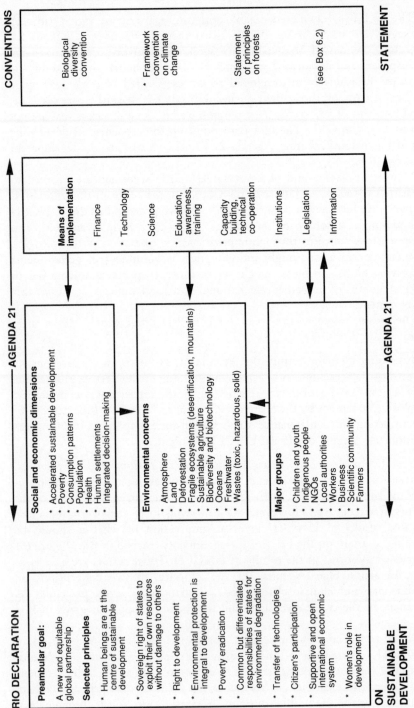

CONVENTIONS

* Biological diversity convention

* Framework convention on climate change

* Statement of principles on forests

(see Box 6.2)

STATEMENT

AGENDA 21

Social and economic dimensions

* Accelerated sustainable development
* Poverty
* Consumption patterns
* Population
* Health
* Human settlements
* Integrated decision-making

Environmental concerns

* Atmosphere
* Land
* Deforestation
* Fragile ecosystems (desertification, mountains)
* Sustainable agriculture
* Biodiversity and biotechnology
* Oceans
* Freshwater
* Wastes (toxic, hazardous, solid)

Major groups

* Children and youth
* Indigenous people
* NGOs
* Local authorities
* Workers
* Business
* Scientific community
* Farmers

Means of implementation

* Finance

* Technology

* Science

* Education, awareness, training

* Capacity building, technical co-operation

* Institutions

* Legislation

* Information

AGENDA 21

RIO DECLARATION

Preambular goal:

A new and equitable global partnership

Selected principles

* Human beings are at the centre of sustainable development

* Sovereign right of states to exploit their own resources without damage to others

* Right to development

* Environmental protection is integral to development

* Poverty eradication

* Common but differentiated responsibilities of states for environmental degradation

* Transfer of technologies

* Citizen's participation

* Supportive and open international economic system

* Women's role in development

ON SUSTAINABLE DEVELOPMENT

Figure 6.1 The results of UNCED (adopted documents)

Box 6.2 UNCED conventions and principles

Two conventions that were negotiated by inter-governmental committees outside UNCED were opened for signature at UNCED.

The *United Nations Framework Convention on Climate Change* calls for the stabilization of greenhouse gas concentrations in the atmosphere at a level that prevents a dangerous anthropogenic interference with the climate system (Article 2). This relatively vague formulation disappointed many countries. It does not provide for a specific timetable such as the initially targeted return to 1990-level emissions by the year 2000, which was then blocked by the USA. However, in a major reversal of policy, the new President of the USA committed the United States to reducing emissions to 1990 levels by the year 2000. The Convention does reflect a global consensus that climate change is a problem that needs to be addressed by all countries and sets up a system of monitoring national greenhouse gas emissions (Article 4a).

The *Convention on Biological Diversity* is to curb the destruction of biological species, habitats and ecosystems. The most important provisions are:

- adoption of national regulations for the conservation of biological resources;
- legal responsibility of governments for transnational impacts of their private corporations;
- assistance to developing countries from the GEF (see Box 6.1);
- transfer of technology without prejudice to intellectual property rights;
- regulation of biotechnology firms;
- access to and ownership of genetic material;
- compensation to developing countries for extraction of their genetic materials.

The USA did not sign the Convention, arguing that it would unduly restrict their biotechnology industry, but again reversed its position and signed the Convention on 4 June 1993.

The *Non-Legally Binding Authoritative Statement of Principles for a Global Consensus on the Management, Conservation and Sustainable Development of all Types of Forests* is the result of disagreement on an initially envisaged Convention on Forests. It has been adopted as an annex to Agenda 21. Opinions on its value differ profoundly, ranging from 'worse than useless – a charter for deforestation' (Greenpeace summary of 7 July 1992) to the 'first global consensus on the management, conservation and sustainable development of forests – a good basis for future policy' (K. Töpfer, Federal Minister for the Environment, Nature Conservation and Nuclear Safety, Germany, in *Earth Summit Times*, 14 September 1992).

---

the Declaration does proclaim the goal of establishing a 'new and equitable partnership' among states by means of new levels of co-operation. It also pronounces a number of agreed principles (see Fig. 6.1) that are taken up in more detail in the Agenda 21.

The Agenda 21 has also been criticized for leaving member states a lot of latitude in interpreting their obligations by frequent qualifications such as 'as appropriate' or 'as far as possible'. Nonetheless, the voluminous list of programmes of the Agenda represents a consensus on what are the main environmental concerns and what action needs to be taken on them, taking account of their interaction with socio-economic goals and priorities. Frequent references have already been made above to different stipulations of Agenda 21, notably regarding the social and economic dimensions of sustainable development and some of the means of its implementation.

It is impossible, of course, to do justice here to over 100 programmes and thousands of activities described in this action plan. Figure 6.1 presents only a summary listing of chapter headings grouped under four parts of Agenda 21 with an indication of their interaction. The effectiveness of the implementation of the Agenda will of course depend on the political will and capability of states to translate rhetorics into policy and action. The newly created Commission on Sustainable Development – the main institutional arrangement established by UNCED – will have the task of keeping this will alive by monitoring and prompting governments and international organizations to implement the commitments contained in Agenda 21.

It remains to be seen how successful the Commission will be in reorienting environmental and development policies towards integrative sustainable development. At least for now, UNCED, especially through its summit segment of kings, presidents and prime ministers, has generated a high degree of awareness of environmental and developmental concerns and their interaction. It has also been exemplary in establishing a participatory approach in which the voices of social groups and non-governmental organizations were heard and to a great extent taken into account by member states. The parallel event of the Global Forum brought together more than 400,000 participants from over 11,000 institutions in 171 countries. This reflects a new opening of international governmental negotiations to non-governmental groups. Hopefully, UNCED will thus be the beginning of a process towards global partnership for sustainable development rather than its culmination. There is, however, a risk that the visibility and popularity of this process will diminish when managed at lower levels of decision-making.

6.3 From national to global compacts

National development planning and policies have been characterized above as fragmented and thus ineffective in addressing environmental and socio-economic interdependences. Improved knowledge about these interdependences permits the definition of alternative – integrative – operational concepts and variables and the development of new planning and policy techniques. The question is now how to overcome vested interests and encrusted institutional structures that tend to resist changes in traditionally mandated responsibilities.

From the above description of instruments of integrative policies at national and sub-national levels, a two-pronged approach might be warranted. On one hand, national, economic, social, environmental and demographic policies need to be reformulated and co-ordinated in response to their mutual effects on each other. This can be considered as a response to the WCED's (1987: 310) call for focusing on the policy sources. On the other hand, central (national) policy reform is not enough. Central plans and policies need to be accepted and followed up at the grass roots by communities and individuals. As shown above, this would require release of central power to grass-roots organizations for the implementation of ecodevelopment at local levels. Power-sharing is at the heart of the effective implementation of – decentralized – sustainable development in countries. A national compact has to be reached among national and local organizations with the possible support and mediation of non-governmental ones.

At the international level, the situation is similar. International approaches to environment and development are fragmented among different inter-national organizations dealing with environment, health, agriculture, industry, trade, finance and socio-economic development. Fragmentation has also prevailed among 'sovereign' nations with some possible exceptions in narrowly defined areas, addressed more or less effectively by international conventions (see above, section 6.1 and Box 6.2). The failure of the negotiations on a New International Economic Order, described above in section 1.1, illustrates a situation where nations were unwilling to agree to power-sharing in supranational arrangements for global policy-making.

Only limited progress has been achieved by UNCED in this regard. Despite exhorting global partnership through new levels of international co-operation in its preamble, the Rio Declaration loses no time to emphasize in its second principle the sovereign right of states to exploit their natural resources according to their own policies. The Declaration does, however, add that states have the responsibility to ensure that their activities do not harm the environment outside their jurisdiction. The question is to what extent such statements can be enforced. The key institution created to oversee the implementation of Agenda 21, the Commission on Sustainable Development, does not appear to have such power. The UN General Assembly, at its 47th session in 1992 (resolution 47/191), restricted the functions of the Commission to generally reviewing progress in the imple-mentation of Agenda 21 and 'considering' information provided by govern-ments. In fact, some developing countries have already expressed their dismay at a Commission that might be 'snooping in our backyard' (according to the *Earth Summit Times*, of 14 September 1992: 10).

In the face of interlocked markets and interlocking planetary crises (WCED 1987: 4–8), the concept of the sovereign nation state needs rethinking. The question is if the time has come now to delegate some of the national powers to supranational arrangements and institutions such as a 'World Economic Council', 'Green Security Council' or 'UN Development Security Council' (see Box 6.3). Of course, such power delegation should be free from any coercion, i.e., with the full consent of the 'United' Nations member states. In this sense, 'global partnership' would have to be replaced, or perhaps incorporated, in a global compact on sustainable development. Such a compact should not call for yet another summit meeting (UNDP 1992b: 8). Rather, it should provide for the implementation of the recommendations of the summit in Rio by means of a strong supranational institution. Such an institution should be capable of monitoring critically and enforcing effectively compliance with internationally agreed conventions and action plans. It would have to summon the spirit of social partnership for the expansion of international agreement into all areas of transboundary effects of sustainable development. Such agreement would include, in particular, the acceptance of liability and compensation for transboundary environmental damage (Tolba 1992).

However, in light of the generally perceived sanctity of national sovereignty, the creation of a powerful global institution appears to be utopian. A more realistic approach, at least for now, is to strengthen international co-operation through systematically reinforcing and expanding – to cover social and economic concerns – existing international environmental treaties and conventions. A new international consensus has to be built up for not yet, or not adequately, covered areas such as desertification or tropical forests. As a result, a co-ordinated system of 'governance without government' might emerge (Young, Demko and Ramakrishna 1991: 6).

Much needs to be done now to implement the least-common-denominator recommendations of Rio. More needs to be done, in the long term, to raise

---

Box 6.3 A Development Security Council

As part of a 'visionary approach' to implementing 'human development', the UNDP (1992b: 82) has proposed the establishment of a Development Security Council, similar to the United Nations Security Council. This Council would establish the broad international policy framework for all global development issues, including food and ecological security. It would be composed of 11 permanent members (China, France, Germany, Japan, Russia, the United Kingdom and the USA, joined by the most populous country in each developing region: India, Brazil, Nigeria and Egypt). Its main tasks would be to:

1 design a global policy framework in all key economic and social areas;
2 finance 'development shortfalls' between global development potential and resource availability by means of a global budget; and
3 provide policy co-ordination among international development and financial institutions.

Box 6.4   Closing statements at UNCED

The Rio Conference does not end in Rio.
                    (F. Collor, President of Brazil and President of UNCED)

The road from Rio will be long, exciting, challenging.
                    (M.F. Strong, Secretary-General of UNCED)

Our road is now illuminated by a new light, the spirit of Rio.
          (B. Boutros-Ghali, Secretary-General of the United Nations)

---

the level of the common denominator. Those seem also to be the immediate reactions of the key managers of UNCED at the end of the Conference (see Box 6.4).

REFERENCES

Abdalla, I.S. (1977) 'Development planning reconsidered', in J.J. Nossin (ed.) *Surveys for Development, a Multidisciplinary Approach*, Amsterdam: Elsevier.

Adelman, I. and Fetini, H. (1990) 'Development strategies and the environment', paper given at Conference on the Environment, WIDER, Helsinki, 2–7 September.

Adelman, I., Taft Morris, C., Fetini, H., Golan-Hardy, E. (1992) 'Institutional change, economic development, and the environment', *Ambio* 21, 1: 106–11.

Alfsen, K.H., Bye, T. and Lorentson, L. (1987) *National Resource Accounting and Analysis, the Norwegian Experience 1978–1986*, Oslo: Central Bureau of Statistics.

Archibugi, F. (1992) 'The disciplinary implications of environmental planning and evolution', Working Paper (WP)/92.4a, Planning Studies Centre (Rome).

Archibugi, F., Delors, J. and Holland, S. (1978) 'International crisis & planning for development', in S. Holland (ed.) *Beyond Capitalist Planning*, Oxford (reprint (RP)/83/A/11, chs 9 and 10 by the Centro di studi e piani economici, Rome).

Bartelmus, P. (1979) 'Limits to development – environmental constraints of human needs satisfaction', *Journal of Environmental Management* 9: 255–69.

—— (1980) *Economic Development and the Human Environment – a Study of Impacts and Repercussions with Particular Reference to Kenya*, Munich and London: Weltforum.

—— (1986) *Environment and Development*, Boston, London and Sydney: Allen & Unwin.

—— (1992a) 'Accounting for sustainable growth and development', *Structural Dynamics* 3, 2: 241–60.

—— (1992b) 'Environmental accounting and statistics', *Natural Resources Forum* 16, 1: 77–84.

Bartelmus, P., Lutz, E. and Schweinfest, S. (1992) *Integrated Environmental and Economic Accounting – A Case Study for Papua New Guinea*, Environment Working Paper No. 54, The World Bank.

Bartelmus, P., Stahmer, C. and van Tongeren, J. (1991) 'Integrated environmental and economic accounting: framework for a SNA satellite system', *Review of Income and Wealth* 37, 2: 111–47.

Bartelmus, P. and Tardos, A. (1993) 'Integrated environmental and economic accounting – methods and applications', *Journal of Official Statistics* 9, 1: 179–88.

Bartelmus, P. and van Tongeren, J. (1993) 'Selected issues in integrated environmental-economic accounting', in A. Franz and C. Stahmer (eds) *Approaches to Environmental Accounting, Proceedings of the IARIW Conference on Environmental Accounting, Baden, Austria, 27–29 May 1991*, Heidelberg: Physica-Verlag.

Bartsch, W.H. (1993) 'Integrated human resources planning', in United Nations *Handbook on Integrated Macroeconomic Development Planning and Management for Sustainable Development: Guidelines for Island Developing Countries*, New York: United Nations.

Bhalla, A.S. and Bifani, P. (1992) 'Some global issues', in A.S. Bhalla (ed.) *Environment, Employment and Development*, Geneva: ILO.

Bilsborrow, R.E. and Okoth-Ogendo, H.W.O. (1992) 'Population-driven changes in land use in developing countries', *Ambio* 21, 1: 37–45.

Boo, A. de, Bosch, P., Gorter, C. and Keuning, S. (1993) 'An environmental module and the complete system of national accounts', in A. Franz and C. Stahmer (eds) *Approaches to Environmental Accounting, Proceedings of the IARIW Conference on Environmental Accounting, Baden, Austria, 27–29 May 1991*, Heidelberg: Physica-Verlag.

Boulding, K.E. (1966) 'The economics of the coming spaceship earth', in H. Jarret (ed.) *Environmental Quality in a Growing Economy*, Baltimore: Johns Hopkins Press for Resources for the Future.

Brown, L.R. (1984) 'Securing food supplies', in L.R. Brown *et al. State of the World 1984 – A Worldwatch Institute Report on Progress Toward a Sustainable Society*, New York and London: Norton.

—— (1993) 'A new era unfolds', in L. R. Brown *et al. State of the World 1993 – A Worldwatch Institute Report on Progress Toward a Sustainable Society*, London and New York: Norton.

Brown, L.R. and Young, J.E. (1990) 'Feeding the world in the nineties', in L.R. Brown *et al. State of the World 1988 – A Worldwatch Institute Report on Progress Toward a Sustainable Society*, New York and London: Norton.

Caldwell, L.K. (1972) *In Defense of Earth: International Protection of the Biosphere*, London: Bloomington.

Carson, R. (1965) *Silent Spring*, London: Penguin.

Clark, B.D., Bisset, R. and Walthern, P. (1980) *Environmental Impact Assessment*, London: Mansell.

Corbin, P.B. (1993a) 'Policy and strategy formulation', in United Nations *Integrated Macroeconomic Development Planning and Management for Sustainable Development: Guidelines for Island Developing Countries*, New York: United Nations.

—— (1993b) 'Trust fund formulation: a strategy for sustainable development', in United Nations *Integrated Macroeconomic Development Planning and Management for Sustainable Development: Guidelines for Island Developing Countries*, New York: United Nations.

Costanza, R. (1991) 'The ecological economics of sustainability: investing in natural capital', in R. Goodland, H. Daly, S. El Serafy and B. von Droste (eds) *Environmentally Sustainable Enconomic Development: Building on Brundtland*, Paris: UNESCO.

Costanza, R. (ed.) (1991) *Ecological Economics: the Science and Management of Sustainability*, New York: Columbia University Press.

Curry-Lindahl, K. (1972) *Conservation for Survival: an Ecological Strategy*, New York: Morrow.

Daly, H.E. (1989) 'Toward a measure of sustainable social net national product', in Y.J. Ahmed, S. El Serafy and E. Lutz (eds) *Environmental Accounting for Sustainable Development*, Washington, DC: The World Bank,

—— (1991a) 'Elements of environmental macroeconomics', in R. Costanza (ed.) *Ecological Economics: the Science and Management of Sustainability*, New York: Columbia University Press.

—— (1991b) 'From empty-world to full-world economics: recognizing an historical turning point in economic development', in R. Goodland, H. Daly, S. El Serafy

153

and B. von Droste (eds) *Environmentally Sustainable Economic Development: Building on Brundtland*, Paris: UNESCO.

—— (1991c) 'Sustainable development: from concepts and theory towards operational principles', in K. Davis and M.S. Bernstein (eds) *Resources, Environment and Population: Present Knowledge, Future Options*, New York and Oxford: Oxford University Press.

Daly, H.E. and Cobb Jr, J.B. (1989) *For the Common Good: Redirecting the Economy Towards Community, the Environment, and a Sustainable Future*, Boston, Mass.: Beacon Press.

Dasgupta, P. (1992) 'Population, resources, and poverty', *Ambio* 21, 1: 95–101.

Dasgupta, P. and Mäler, K.-G. (1990) 'The environment and emerging development issues', paper prepared for the World Bank's Annual Conference on Development Economics, Washington, DC, 26–27 April 1990.

Desai, N. (1991) 'A development planner looks at environmental management', in D. Eröcal (ed.) *Environmental Management in Developing Countries*, Paris: OECD.

Devall, B. and Sessions, G. (1985) *Deep Ecology: Living as if Nature Mattered*, Layton, UT: Peregrine Smith.

Doeleman, J.A. (1992) 'Employment concerns and environmental policy', in A.S. Bhalla (ed.) *Environment, Employment and Development*, Geneva: ILO.

Dorfman, R., Samuelson, P. and Solow, R. (1958) *Linear Programming and Economic Analysis for Regional Devlelopment*, New York: The Free Press.

Dower, R.C. and Zimmerman, M.B. (1992) *The Right Climate for Carbon Taxes: Creating Economic Incentives to Protect the Atmosphere*, Washington, DC: World Resources Institute.

Droste (von), B. and Dogsé, P. (1991) 'Sustainable development: the role of investment', in R. Goodland, H. Daly, S. El Serafy and B. von Droste (eds) *Environmentally Sustainable Economic Development: Building on Brundtland*, Paris: UNESCO.

Eckholm, E.P. (1977) *The Picture of Health: Environmental Sources of Disease*, New York: Norton.

El Serafy, S. (1989) 'The proper calculation of income from depletable natural resources', in Y.J. Ahmad, S. El Serafy and E. Lutz (eds) *Environmental Accounting for Sustainable Development*, Washington, DC: The World Bank.

Food and Agriculture Organization of the United Nations (FAO), United Nations Fund for Population Activities (UNFPA) and International Institute for Applied Systems Analysis (IIASA) (1982) *Potential Population Supporting Capacities of Lands in the Developing World*, Rome: FAO.

Frenkiel, F.N. and Goodall, D.W. (eds) (1978) *Simulation Modelling of Environmental Problems*, SCOPE Report 9, Chichester: Wiley.

Gastaldo, S. (1992) 'Approches économiques de l'environnement', internal paper of the Institut National de la Statistique et des Etudes Economiques (Paris).

Giannone, A. and Carlucci, M. (1991) 'Environmental variables in national accounts – a case study for Italy', paper submitted to the Special Conference on Environmental Accounting of the International Association for Research in Income and Wealth, Baden, 27–29 May.

Goldsmith, E. *et al.* (1972) *Blueprint for Survival*, Boston: Houghton Mifflin.

Goodland, R. (1991) 'The case that the world has reached limits: more precisely that current throughput growth in the global economy cannot be sustained', in R. Goodland, H. Daly, S. El Serafy and B. von Droste (eds) *Environmentally Sustainable Economic Development: Building on Brundtland*, Paris: UNESCO.

Gore, A. (1993) *Earth in the Balance – Ecology and the Human Spirit*, New York: Plume.

Government of Canada (1990) *Canada's 'Green Plan'*, Ottowa.

—— (1991) *Canada's National Report – United Nations Conference on Environment and Development, Brazil, June 1992*, Ottawa.

Government of Japan (1991) *Environment and Development, Japan's Experience and Achievement – Japan's National Report to UNCED 1992*.

Grambsch, A.E. and Michaels, R.G. with Peskin, H.M. (1993) 'Taking stock of nature: environmental accounting for the Chesapeake Bay', in E. Lutz (ed.) *Toward Improved Accounting for the Environment, an UNSTAT-World Bank Symposium*, Washington, DC: The World Bank.

Hartwick, J.M. and Hageman, A.P. (1991) 'Economic depreciation of mineral stocks and the contribution of El Serafy', Report for the World Bank, July 1991.

Hicks, J.R. (1946) *Value and Capital*, 2nd edn, Oxford: Oxford University Press.

Holdgate, M.W., Kassas, M. and White, G.F. (eds) (1982) *The World Environment 1972–1982: a Report by the United Nations Environment Programme*, Dublin: Tycooly.

Howarth, R.B. and Norgaard, R.B. (1990) 'Intergenerational resource rights, efficiency, and social optimality', *Land Economics* 66, 1: 1–11.

Hueting, R. (1989) 'Correcting national income for environmental losses: towards a practical solution', in Y.J. Ahmad, S. El Serafy and E. Lutz (eds) *Environmental Accounting for Sustainable Development*, Washington, DC: The World Bank.

ILO (International Labour Office) (1977) *Employment, Growth and Basic Needs: a One-World Problem*, New York: Praeger.

Institut National de la Statistique et de Etudes Economiques (INSEE) (1986a) *Les Comptes du Patrimoine Naturel*, les collections de l'inséé, 137/138C.

—— (1986b) *Les Comptes Satellites de l'Environnement, Méthodes et Résultats*, les collections de l'inséé, 130C.

Inter-governmental Panel on Climate Change (IPCC) (1990) *IPCC First Assessment Report – Overview and Policy-maker Summaries*, WHO and UNEP.

Inter-Secretariat Working Group on National Accounts (1994) *System of National Accounts 1993* (United Nations publications – sales no. E.94.XVII.4), Brussels/ Luxembourg, New York, Paris, Washington, DC.

Isard, W., Choguill, C.L., Kissin, J., Seyfarth, R.H., Tatlock, R., with the assistance of Basset, K.E., Furtado, J.C. and Izumita, R.M. (1972) *Ecologic-Economic Analysis for Regional Development*, New York: The Free Press.

Islam, N. and Henault, G.M. (1979) 'From GNP to basic needs: a critical review of development and development administration', *International Review of Administrative Sciences* 45, 3: 253–67.

Jacobson, J.L. (1988) *Environmental Refugees: a Yardstick of Habitability*, World Watch Paper 86, Washington, DC: World Watch Institute.

—— (1993) 'Closing the gender gap in development', in L.R. Brown *et al. State of the World 1993 – A Worldwatch Institute Report on Progress Toward a Sustainable Society*, New York and London: Norton,

Jänicke, M., Mönch, H., Ranneberg, T. and Simonis, U.E. (1989) 'Structural change and environmental impact', *Intereconomics*, Jan.–Feb.: 24–35.

Jayawardena, L. (1991) *A Global Environmental Compact for Sustainable Development: Resource Requirements and Mechanisms*, Helsinki: WIDER.

Karshenas, M. (1992) 'Environment, development and employment: some conceptual issues', in A.S. Bhalla (ed.) *Environment, Employment and Development*, Geneva: ILO.

Kothari, R. (1980) 'Environment and alternative development', *Alternatives* 5, 4: 427–75.

Lanly, J.-P. (1982) *Les Ressources Forestières Tropicales*, Etude FAO: Forêts, Rome: FAO.

Leipert, C. (1984) 'Bruttosozialprodukt, defensive Ausgaben und Nettowohlfahrts-messung. Zur Ermittlung eines von Wachstumskosten bereinigten Konsumindi-kators', *Zeitschrift für Umweltpolitik* 3: 229–55.

—— (1985) 'Social costs as a factor of economic growth', paper for the Conference for a New Economics: The Other Economic Summit, London, 16–19 April 1985.

—— (1987) 'A critical appraisal of gross national product, the measurement of net national welfare and environmental accounting', *Journal of Economic Issues* 21, 1: 357–73.

—— (1989) 'National income and economic growth: the conceptual side of defensive expenditures', *Journal of Economic Issues* 23, 3: 843–56.

Leontief, W. (1970) *Environmental Repercussions and the Economic Structure: an Input-Output Approach*, Review of Economics and Statistics 52.

Loraine, J.A.C. (1972) *The Death of Tomorrow*, London: Heinemann.

Lutz, E. and Munasinghe, M. (1993) 'Integration of economic concerns into economic analyses with special emphasis on valuation issues', in E. Lutz (ed.) *Toward Improved Accounting for the Environment, an UNSTAT-World Bank Symposium*, Washington, DC: The World Bank.

McNeely, J.A. *et al.* (1990) *Conserving the World's Biological Diversity*, Gland: IUCN.

Mäler, K.G. (1991) 'National accounts and environmental resources', *Environmental and Resource Economics* 1: 1–15.

Meadows, D.H., Meadows, D.L., Randers, J. and Behrens III, W.W. (1972) *The Limits to Growth*, New York: Universe Books.

Morrison, W.I. (1993) 'Data and modelling', in United Nations *Integrated Macro-economics Development Planning and Management for Sustainable Development: Guidelines for Island Developing Countries*, New York: United Nations.

Myers, N. (1992) 'Population/environment linkages: discontinuities ahead', *Ambio* 21, 1: 116–18.

Naess, A. (1976) 'The shallow and the deep, long-range ecology movement, a summary', *Inquiry* 16: 95–100.

National Institute of Public Health and Environmental Protection (1992) *National Environmental Outlook 2, 1990–2010*, Bilthoven: RIVM.

Newels, R. (1993) 'Spatial planning and management', in United Nations *Integrated Macroeconomic Development Planning and Management for Sustainable Develop-ment: Guidelines for Island Developing Countries*, New York: United Nations.

Nordhaus, W.D. and Tobin, J. (1973) 'Is growth obsolete?', *Studies in Income and Wealth* 38: 509–64.

Odum, E.P. (1971) *Fundamentals of Ecology*, 3rd edn, Philadelphia: W.B. Saunders.

Ohlin, G. (1992) 'The population concern', *Ambio* 21, 1: 6–9.

Opschoor, J.B. (1989) 'North-south trade, resource degradation and economic security', *Bulletin of Peace Proposals* 20, 2: 135–42.

Organization for Economic Co-operation and Development (OECD) (1973) *List of Social Concerns Common to Most OECD Countries*, Paris: OECD.

—— (1989) *Economic Instruments for Environmental Protection*, Paris: OECD.

—— (1991) *Environmental Indicators: a Preliminary Set*, Paris: OECD.

—— (1992) *Science Responds to Environmental Threats*, Paris: OECD.

Organization for Eastern Caribbean States (OECS), Natural Resources Management Unit (1989) *Simulation Model on Sustainable Development*, St Lucia: OECS.

Organization of American States (OAS) (1984) *Integrated Regional Development Planning: Guidelines and Case Studies from OAS Experience*, Washington, DC.: OAS.

Otterbein: K. (1991) 'Mega-Städte, Mega-Krisen', *Dritte Welt Presse* 8, 1: 1–2.

Panayotou, T. (1991) 'Economic incentives in environmental management and their irrelevance to developing countries', in D. Eröcal (ed.) *Environmental Management in Developing Countries*, Paris: OECD.

Parliamentarians for Global Action, International Debt and Economic Reform Programme (1990) 'Prospects for development in the 1990s', discussion paper for the Parliamentary Workshop and Parliamentarian-Diplomat Dialogue, April 1990.

Pearce, D., Markandya, A. and Barbier, E.B. (1990) *Blueprint for a Green Economy*, London: Earthscan Publications Ltd.

Pearce, D. and Turner, R. (1990) *Economics of Natural Resources and the Environment*, Baltimore: The Johns Hopkins University Press.

Peskin, H.M. (1989) 'A proposed environmental accounts framework', in Y.J. Ahmad, S. El Serafy and E. Lutz (eds) *Environmental Accounting for Sustainable Development*, Washington, DC: The World Bank.

Pezzey, J. (1989) *Economic Analysis of Sustainable Growth and Sustainable Development*, Environment Department Working Paper No. 15, Washington, DC: The World Bank.

Postel, S. (1990) 'Toward a New "Eco"-nomics', *World-Watch* 3, 5: 20–8.

—— (1992) 'Denial in the decisive decade', in L.R. Brown *et al. State of the World 1992 – A Worldwatch Institute Report on Progress Toward a Sustainable Society*, New York and London: Norton.

Pronk, J. and Haq, M. (1992) *The Hague Report, Sustainable Development – from Concept to Action*.

Rasmuson, M. and Zetterström, R. (1992) 'World population, environment and energy demands', *Ambio* 21, 1: 70–4.

Repetto, R., Dower, R.C., Jenkins, R. and Geoghegan, F. (1992) 'Green fees: how a tax shift can work for the environment and the economy', WRI Publications Brief.

Repetto, R., Magrath, W., Wells, M., Beer, C. and Rossini, F. (1989) *Wasting Assets, Natural Resources in the National Income Accounts*, Washington, DC: World Resources Institute.

Rodgers, G. (1984) *Poverty and Population: Approaches and Evidence*, Geneva: ILO.

Rubenstein, D. (1989) 'Black oil, red ink', *CA Magazine*, Nov.: 30–5.

Ryan, J.C. (1992) 'Conserving biological diversity' in L.R. Brown *et al. State of the World 1992 – A Worldwatch Institute Report on Progress Toward a Sustainable Society*, New York and London: Norton,

Sachs, I. (1976) 'Environment and styles of development', in W.H. Matthews (ed.) *Outer Limits and Human Needs*, Uppsala: Dag Hammarskjöld Foundation.

—— (1980) *Stratégies de l'Ecodéveloppement*, Paris: Editions Ouvrières.

Samuelson, P.A. and Nordhaus, W.D. (1985) *Economics*, 12th edn, New York and London: McGraw-Hill.

—— (1992) *Economics*, 14th edn, New York: McGraw-Hill.

Schäfer, D. and Stahmer, C. (1990) 'Conceptual considerations on satellite systems', *Review of Income and Wealth* 36, 2: 167–76.

Schmidheiny, S. (1992) 'The business of business: eco-efficiency', *Earth Summit Times*, 27 April 1992: 7.

Seers, D. (1983) *The Political Economy of Nationalism*, London: Oxford University Press,

Shaw, R.P. (1989) 'Rapid population growth and environmental degradation: ultimate *versus* proximate factors', *Environmental Conservation* 16, 3: 199–208.

Solórzano, R., de Camino, R., Woodward, R., Tosi, J., Watson, V., Vásquez, A., Villalobos, C., Jiménez, J., Repetto, R. and Cruz, W. (1991) *Accounts Overdue: Natural Resource Depreciation in Costa Rica*, San José: Tropical Science Center and Washington, DC: World Resources Institute.

Solow, R. (1992) *An Almost Practical Step Toward Sustainability*, Washington, DC: Resources for the Future.

Stöhr, W.B. (1981) 'Development from below: the bottom-up and periphery-inward development paradigm', in W.B. Stöhr and D.R.F. Taylor (eds) *Development from Above or Below: the Dialectics of Regional Planning in Developing Countries*, New York: Wiley.

Theys, J. (1989) 'Environmental accounting in development policy: the French experience', in Y.J. Ahmad, S. El Serafy and E. Lutz (eds) *Environmental Accounting for Sustainable Development*, Washington, DC: The World Bank.

Todaro, M.P. (1977) *Economic Development in the Third World: an Introduction to Problems and Policies in a Global Perspective*, London: Longman.

Tolba, M.K. (1992) 'Sovereignty and the global environment', remarks at the Massachusetts Institute of Technology, 24 September 1992, UNEP series of speeches.

ul Haq, M. (1989) 'Human dimension in development', *Journal of Development Planning* 19: 249–58.

United Nations (1973) *Report of the United Nations Conference on the Human Environment*, Stockholm 5–16 June 1972 (sales no. E.73.II.A.14), New York: United Nations.

—— (1976) *Report of Habitat: United Nations Conference on Human Settlements*, New York: United Nations.

—— (1977) *Provisional International Guidelines on the National and Sectoral Balance Sheet and Reconciliation Accounts of the System of National Accounts* (sales no. E.77.XVII.10), New York: United Nations.

—— (1978) *Water Development and Management: Proceedings of the United Nations Water Conference Mar del Plata, Argentina, March 1977*, Oxford: Pergamon.

—— (1982) *Towards the New International Economic Order: Report of the Director General for Development and International Economic Co-operation*, New York: United Nations.

—— (1984) *A Framework for the Development of Environment Statistics* (sales no. E.84.XVII.12), New York: United Nations.

—— (1988) *Concepts and Methods of Environment Statistics: Human Settlements Statistics – a Technical Report* (sales no. E.88.XVII.14), New York: United Nations.

—— (1991a) *Concepts and Methods of Environment Statistics: Statistics of the Natural Environment – a Technical Report* (sales no. E.91.XVII.18), New York: United Nations.

—— (1991b) *Report of the Ad Hoc Committee of the Whole for the Preparation of the International Development Strategy for the Fourth United Nations Development Decade* (A/45/41), New York: United Nations.

—— (1992a) *Environmental Accounting, Current Issues, Abstracts and Bibliography* (sales no. E.92.II.A.23), New York: United Nations.

—— (1992b) *Environmentally Sound Technology for Sustainable Development* (sales no. E.92.II.A.6), New York: United Nations.

—— (1992c) *1990 Energy Statistics Yearbook* (sales no. E/F.92.XVII.3), New York: United Nations.

—— (1992d) *Revolution and Reform in Eastern Europe and the Soviet Union: the Global Development Impact* (the CDP Report 1991), New York: United Nations.

—— (1992e) 'Structural change in the world economy, energy use and the environment' (DESD/DPAD/1992/1, mimeographed).

—— (1993a) *Integrated Environmental and Economic Accounting, interim version* (sales no. E.93.XVII.12), New York: United Nations.

—— (1993b) *Trends in International Distribution of Gross World Product* (sales no. E.92.XVII.7), New York: United Nations.

—— (1993c) *Report of the United Nations Conference on Environment and Development*, Rio de Janeiro, 3–14 June 1992 vol. 1, Resolutions adopted by the Conference (sales no. E.93.I.8, vol. 1), New York: United Nations.

United Nations Centre for Human Settlements (UNCHS) (1990) *The Global Strategy for Shelter to the Year 2000*, Nairobi: UNCHS.

—— (undated) *People, Settlements, Environment and Development*, Nairobi: UNCHS.

United Nations Centre on Transnational Corporations (UNCTC) (1991) 'Options to facilitate transfer of environmentally sound technologies to developing countries in favourable terms' (draft papers, packet 1).

United Nations Conference on the Human Environment (1972) *Development and Environment*, Report and Working Papers of a Panel of Experts, Mouton: United Nations and Ecole Pratique des Hautes Etudes.

United Nations Development Programme (UNDP) (1991) *Human Development Report 1991*, New York and Oxford: Oxford University Press.

—— (1992a) *Handbook and Guidelines for Environmental Management and Sustainable Development*, New York: UNDP.

—— (1992b) *Human Development Report 1992*, New York and Oxford: Oxford University Press.

United Nations Development Programme (UNDP), United Nations Environment Programme (UNEP) and the World Bank (1991) *Global Environmental Facility*.

United Nations Educational, Scientific and Cultural Organization (UNESCO) (1981) *International Co-ordinating Council of the Programme on Man and the Biosphere (MAB)*, MAB report series, no. 53, Paris: UNESCO.

United Nations Environment Programme (UNEP) (1975) 'The proposed programme' (UNEP/GC/30), Nairobi.

—— (1978) *Review of the Areas Environment and Development and Environmental Management*, Nairobi: UNEP.

—— (1992) *Saving Our Planet, Challenge and Hopes, The State of the Environment (1972–1992)*, Nairobi: UNEP.

United Nations Environment Programme (UNEP) and United Nations Conference on Trade and Development (UNCTAD) (1974) Symposium on Patterns of Resource Use, Environment and Development Strategy, 'The Cocoyoc Declaration' (mimeographed).

United Nations Population Fund (UNFPA) (1991) *Population and the Environment: the Challenges Ahead*, London: Banson.

US Department of Health, Education and Welfare (1961) *Toward a Social Report*, Washington, DC.

van Tongeren, J., Schweinfest, S., Lutz, E., Gomez Luna, M. and Guillen Martin, F. (1991) *Integrated Environmental and Economic Accounting – A Case Study for Mexico*, Environment Working Paper No. 50, Washington, DC: The World Bank.

Vitousek, P.M., Ehrlich, P.R., Ehrlich, A.H. and Matson, P.A. (1986) 'Human appropriation of the products of photosynthesis', *Bioscience* 36, 6: 368–73.

Vivian, J.M. (1991) *Greening at the Grassroots: People's Participation in Sustainable Development*, UNRISD Discussion Paper No. 22, Geneva: UNRISD.

Walras, L. (1926) *Elements of Pure Economics, or the Theory of Social Wealth*, London: Allen & Unwin 1954.

Walters, C.J. (1978) 'Obergurgl: development in high mountain regions of Austria', in C.S. Holling (ed.) *Adaptive Environmental Assessment and Management*, Chichester: Wiley.

World Bank (1985) *Desertification in the Sahelian and Sudanian Zones of West Africa*, Washington, DC: The World Bank.
—— (1990) 'Papua New Guinea, the forestry sector: a tropical forestry action plan review', a report prepared by the World Bank under the auspices of the Tropical Forestry Action Plan.
—— (1992) *World Development Report 1992 – Development and the Environment*, Oxford and New York: Oxford University Press.
World Commission on Environment and Development (WCED) (1987) *Our Common Future*, Oxford and New York: Oxford University Press.
World Food Council (1991) *Hunger and Malnutrition in the World* (WFC/1991/2), Rome: World Food Council.
World Resources Institute (WRI) (1992) *World Resources 1992–93*, New York and Oxford: Oxford University Press.
Young, D.R., Demko, G.J. and Ramakrishna, K. (1991) *Global Environmental Change and International Governance*, summary and recommendation of a conference held at Dartmouth College, Hannover, NH, June 1991.
Young, M.D. (1991) 'Natural resource accounting: some Austrialian experiences and observations', paper submitted to the Special Conference on Environmental Accounting of the International Association for Research in Income and Wealth, Baden, 27–29 May.
Zylicz, T. (1991) 'The role for economic incentives in international allocation of abatement effort', in R. Costanza (ed.) *Ecological Economics: the Science and Management of Sustainability*, New York: Columbia University Press.

INDEX

(text box numbers in italics)

activity analysis 73–4, *3.3*, 78, 121
Agenda 21 79, 147–8
air pollution 21, 26

basic human needs 6–7, 113–15
biodiversity 19
biomass appropriation 17

Canada *see* Green Plan, Canada
capital: accumulation 42, *2.3*, 90; human 63, 90, 94 (*see also* human resource development); institutional 63; natural, maintenance 40, 63 (*see also* natural assets); produced 36, 63; productivity 57
carbon tax 104, *4.9*
carrying capacity 17, 68–9, 112, 122
central planning 78, 124, 126
Chesapeake Bay 58
climate change *see* global warming
Coase theorem 139
command-and-control measures 101–2
Commission on Sustainable Development 148, 149
consumption patterns 21–2, 88–9
corporate environmental accounting 50–1, *2.7*
cost-benefit analysis *2.6*, 129
Costa Rica *2.9*, *4.7*

debt-for-nature swaps 96, *4.7*
deep ecology 67, 76–7, *3.4*
defensive expenditures 38–9, *2.1*, 53
deforestation 19–20, 24
depreciation 40, 46, 63

developing countries 3–5
development 1; local *see* ecodevelopment; objectives 1–2, *1.1*; planning 107–8, *5.1*, 126; projects 128–9; security council 150, *6.3*; strategies 2, *1.2*, 108
discount rate 97, 129
distribution of income and wealth *see* equity

Earth Summit *see* United Nations Conference on Environment and Development
ecodevelopment 115–16; definition 119; features 120; models 121–3; strategies, PNG 138–9
economic growth 2–3; and environment 87–8, *4.3*; limits 5, 10, 12, 60, 87–8, 106; socially necessary 115, 143; zero 5, 10
economic instruments 101–5, *4.8*; international 141
ecopreneurship 99
ecoregion 119
ecosystem 81, 116–18, *5.4*; social 120
ecotechniques 120
employment: and environmental protection 94–6; planning 109
empowerment of local populations 124
energy: consumption 20, 25; subsidies 99, 118; throughput 118
environment and development 5, 7, 85
environment statistics 132, *5.8*
environmental accounting *see* System of Integrated Environmental and

Economic Accounting *or* corporate environmental accounting
environmental assets *see* natural assets
environmental concerns *see* environmental issues
environmental costs 36, 44; imputed 50; internalization 50, 52, 99–101
environmental diseases 26
environmental doomsday 5, 10
environmental education 109–10, 133
environmental impact assessment 128–9, *5.6*
environmental indicators 27, *1.8*, *5.8*
environmental issues 10–13, *1.6*
environmental legislation 127–8
environmental plan 127
environmental protection *see* defensive expenditures
environmental refugees 19
environmental security 142–3
environmental standards 27, 48, 72
environmental trust fund 90, 134
environmental-economic exchange system 31–3
Environmentally-adjusted National Income 45, 49, 65
Environmentally-adjusted net Domestic Product 42–4, *2.4*
environmentally sound technology: definition 130; transfer 99, 130–1, *5.7*
equity 67, 71, 97, 120; intergenerational 111; international 143
externalities 32, 75

feasible development *see* sustainable development; *see also* activity analysis
financial resources, new and additional 133; international 143, *6.1*, 144–5
fiscal incentives 52, 100; *see also* economic instruments
food: production 23–4; security 113
full-cost pricing 91, 99

general equilibrium *4.2*, 100
Germany 39
global commons 140
global compact on sustainable development 150
global warming 17–18
Green Plan, Canada 135–7, *5.9*
greenhouse gases *see* global warming

human development 67, 71–2, 110–11; index 4, 68
Human Freedom Index 68, *3.2*
human health 110, 113, *5.2*
human resource development 108–10
human settlements 25–6, 114, *5.3*

income 61; distribution *see* equity; national 61, 63, 65
Index of Sustainable Economic Welfare 54
Indonesia *2.9*
inflation, environmental effects 97
information overload 27
input-output analysis 73, 89, *4.4*, 121, 141
interdependences 73; economic-ecological 29, *1.9*, 114
international conventions 142, *6.2*
investment strategies, environmentally sound 89–90

Japan 126

land degradation 19, 23–4
least developed countries 5
lifestyles *see* consumption patterns

macroeconomic management *4.2*, 87
market valuation 45–7, 50
Mexico 55–7

national accounts 34; satellite 35, 39; system of 35
national compact 149
natural assets 40–1, *2.2*
natural disasters 25; accounting *2.3*
natural patrimony *see* natural assets
natural resources *2.2*; discoveries *2.3*
Net Economic Welfare 53, *2.8*
New International Economic Order 6–7, *1.3*

oceans 21
optimality 60, 75–8; and feasible development 77–8; of natural capital stocks 76; of pollution 100–1; and sustainability 76–8
ozone layer 19

Papua New Guinea 55–7, 92, *4.5*; *see also* sustainable development, national action plan, PNG

partnership: global 147, 150; social 131
policy: framework *see* sustainable
    development, policy framework;
    proactive 84; reactive 85; short-term
    stabilization 86–7
polluter-pays principle 32, 100, 101
pollution: of affluence 11–12; of poverty
    6, 11
population: and environment 111–13;
    indigenous 115; planning 109; *see also*
    carrying capacity
poverty 22–3, *1.7*, 113; eradication
    110–11
precautionary principle 129

quality of life 66

recycling 32
regional planning 124–5, *5.5*
Rio Declaration 145–7, 149

social accounting matrix 68
social indicators 66, *3.1*
sovereignty 140, 149
steady-state economy 106
structural adjustment 87, 96
structural distortion of the economy
    95–6, 98–9
sustainability 61; ecological 64;
    economic 64; factors 64–5, 66;
    indicators 12, 14–16, 60; social 64;
    strong 65; weak 65
sustainable development: definition 69,
    73, 76; indicators 133; national action
    plan, PNG *5.10*, *5.11*, 137–9;
    objectives 7–8, *1.4*; policy directions
    79–80, *4.1*; policies 84–5; policy
    framework 81–4
sustainable economic growth: definition
    70; policies 87–97

sustainable livelihood 110
sustainable national income 38
synergisms 73
System of Integrated Environmental and
    Economic Accounting: case studies
    55–7; framework 37–8; objectives
    35–8

technical assistance 134
trade: barriers 91; and environment
    91–4; GATT obligations and
    environmental exceptions 93, *4.6*;
    international negotiations 93;
    liberalization 91, 141; sanctions 93
trade-offs 73
transfer pricing 92, *4.5*

uncertainty 76, 129
United Nations: Conference on
    Environment and Development 8,
    144–8, *6.2*; Conference on the
    Human Environment 6, 126;
    international development strategies
    see development, strategies
urbanization 25–6
USA *2.8*, 54, 59, 104, *4.9*
user-pays principle 32, 100
utility 32, 75

valuation of natural assets: contingent
    48–9, *2.6*, 52, 57–8; maintenance-cost
    47–8, 51–2; net-price 46–7, *2.5*, 57;
    user-cost 46–7, *2.5*, 57, 99; *see also*
    market valuation

waste dumping 94
water 24
welfare 32, 61, 62; effects 70–1; function
    76
women and environment 67, 121